LEGUMES OF THE ILHA DE MARACÁ

LEGUMES OF THE ILHA DE MARACÁ

G.P. LEWIS & P.E. OWEN

Illustrated by Angela Beaumont

ROYAL BOTANIC GARDENS
KEW

First published 1989

General editor M.J.E. Coode

ISBN 0 947643 15 X

Printed and Bound in Great Britain by
Whitstable Litho Ltd., Whitstable, Kent

PREFACE

The Amazon Basin contains sixty per cent of the world's surviving rainforests, and the majority of this is in Brazil. Every year some 11 million hectares (27 million acres or the combined area of Scotland and Wales) of Amazon rainforest is destroyed together with the immense wealth of creatures that live in it.

The vast majority of rainforest species has not been fully studied and many are still unknown to science. The few species that *are* exploited by man yield a wealth of foods, timbers, essential oils and pharmaceuticals. A quarter of our prescription drugs are of biological origin, and many come from tropical forests. But there are undoubtedly many more lifesaving drugs awaiting discovery. The recent isolation of the alkaloid castanospermine (currently undergoing tests as a potential anti-AIDS drug) from the legume *Alexa canaracunensis* is just such a case discussed in this book.

The Maracá Rainforest Project supported one of the largest scientific teams ever to have worked in the Amazon Basin. I hope that the results of their research will help us to better understand the complexities of the Amazon's highly evolved biota and add support to the need to conserve this very valuable environment with its finely tuned ecology.

This volume amply demonstrates the benefits of undertaking detailed specialist collecting in small, well defined areas of the tropics. It contributes to highlighting the richness of the Amazon flora and shows the importance of tropical rainforests as a storehouse of biological diversity.

John Hemming
Director and Secretary of the Royal Geographical Society

This work is dedicated to the memory of five great men without whose collections and publications our research would have been impossible. They are George Bentham, Adolfo Ducke, Richard and Robert Schomburgk and Richard Spruce. It is also for Taciana Cavalcanti with affection.

CONTENTS

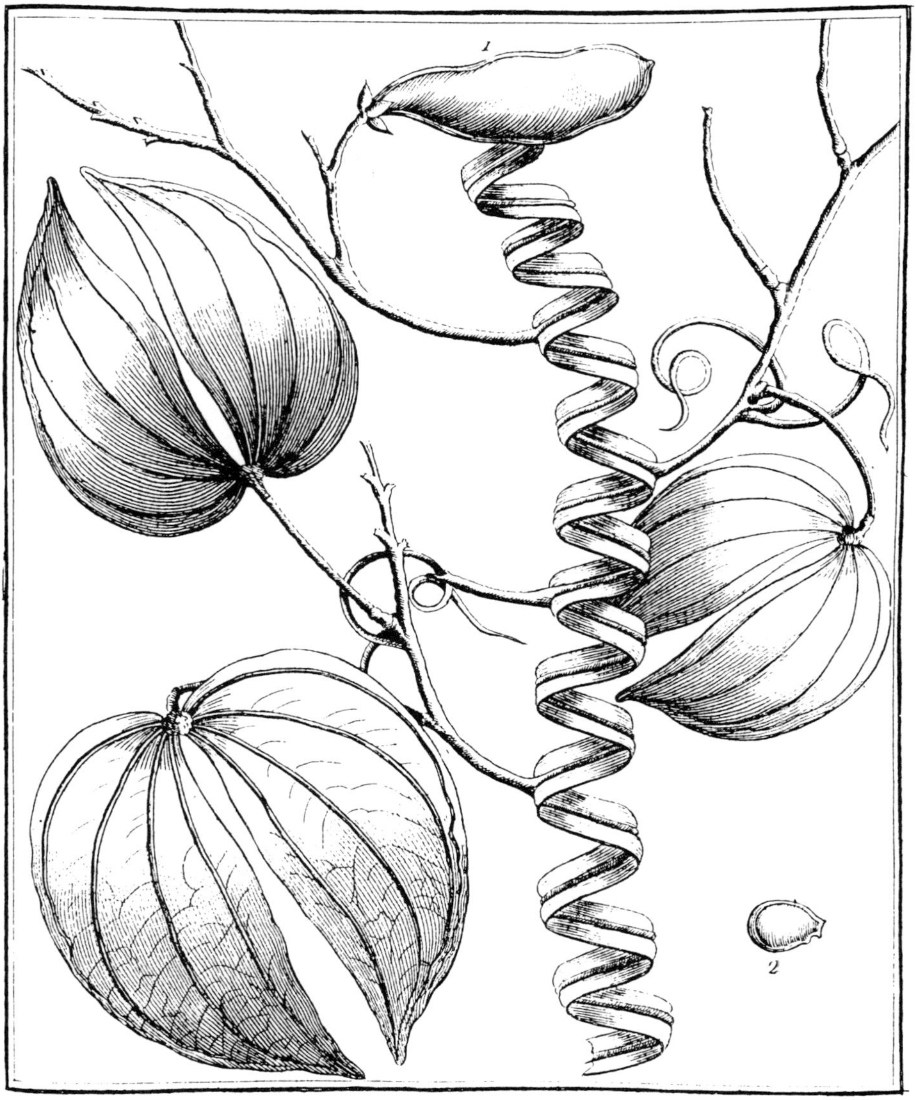

Bauhinia guianensis

ACKNOWLEDGEMENTS

We are delighted to be able to recognise in print the assistance and encouragement given to us by several friends and colleagues during both the field work phase and the following research and write-up resulting in this volume. We trust that nobody is missing from the list that follows.

Amongst the many who deserve special thanks, William Milliken undoubtedly stands out as someone without whom we could not have produced this book. Many of the legumes cited in the text were collected by him and G.P.L. accompanied him on a circumnavigation of Maracá in October 1987 adding significantly to the number of legume species collected from the island. He was responsible for sorting and freighting much of the botanical material to Kew and, with the help of collectors from INPA, went out of his way to collect some large leguminous trees which might otherwise have remained unidentified.

We would like to thank Dr John Hemming, the Director and Secretary of the Royal Geographical Society, for inviting Kew to participate in the Maracá Project and for providing first rate logistical support in the field. Amongst Dr Hemming's team of hard working administrators G.P.L. acknowledges the magnificent support of Steve Bowles, Fiona Watson, Sarah Latham, Sue Whinney and Daphne Hanbury. Life on Maracá would not have been the same without Jill Thompson and Duncan Scott, two good friends. Amazonas, the field station caretaker and Guttemberg Moreno de Oliveira, SEMA's island administrator, deserve special praise for all their hard work throughout the fifteen months of the expedition.

Three other collectors from Kew were on Maracá between February and April 1987 and, as well as providing companionship during that period, have followed the development of this volume with keen interest. They are Raymond Harley, Brian Stannard and Peter Edwards.

Several Brazilians doubled as trail cutters and boatmen and G.P.L. owes much to the survival instincts of Luiz, Antonio and Chico and also to three exceptional people, the Maiongong Indians Vitor, Marco and Luiz.

Dr Jimmy Ratter has continually encouraged us since the inception of this work and provided much of the material cited herein. His amusing anecdotes about Brazilian plant exploration brightened every day on Maracá.

Laura Maria de Moraes Mourão spent some time on the island assisting with the collection of some legumes and provided stimulating debate about aspects of Brazilian ecology and conservation.

Robert Miller provided valuable comment about the ecology of *Hymenaea* species and shared observations of seedling development of *Alexa canaracunensis*.

Gren Lucas, the keeper of the Herbarium at Kew provided funds for G.P.L. during his three visits to the island and sent out replacement camera equipment after a boating accident on the Furo de Maracá in April 1987. He also secured funds for P.E.O. to work at Kew during his one year sandwich studentship and we gratefully acknowledge all this support. We also acknowledge Marlene Freitas da Silva's role as Kew's botanical 'responsavel' during the expedition and Angelo dos Santos' general encouragement and diplomacy. We are grateful to Dr Ghillean T. Prance, formally Director of the New York Botanical Garden and Patricia K. Holmgren, the Herbarium Curator, for providing critical specimens on loan to Kew. Mike Hopkins provided a computer-generated checklist permitting selection of this loan.

Under a great deal of pressure from other work commitments and with a two year study visit to South Africa imminent, Angela Beaumont willingly agreed to prepare the seven black and white fruit figures and design the book cover. We are more than pleased with the results. Thanks are due to Milton Andrews for preparing Map 2.

When the going became tough Barbara MacKinder protected us from other work commitments allowing us to complete the volume on time; she also helped P.E.O. with the preparation of some of the species descriptions and together with Joan Barham tested the workability of the key to genera.

Mark Coode, Roger Polhill, Charles Stirton and Bernard Verdcourt provided continual advice and encouragement; Barry Bolton named the ants inhabiting *Alexa canaracunensis;* Rupert Barneby provided expert advice on *Mimosa* and *Albizia* species; Maria das Lourdes Rico Arce assisted with the identification of some mimosoid genera and Don Windler provided comment on *Crotalaria velutina;* André M. de Carvalho gave advice about *Dalbergia riedelii* and prepared the portuguese introduction, and Haroldo Cavalcante de Lima identified *Hymenolobium petraeum* from fragments brought down with a 10-bore rifle.

John Stone designed the layout of the cover; Tudor Harwood, Milan Svanderlik and Andrew McRobb advised on colour plate production. The text was typed in the Typing Pool and set by Pam Arnold, Brenda Carey, and Margaret Newman.

NEW COMBINATIONS

Two new combinations are proposed in this volume

Mimosoideae:

Albizia corymbosa (L.C. Rich.) G.P. Lewis and P.E. Owen — page 40
Albizia glabripetala (H.S. Irwin) G.P. Lewis and P.E. Owen — page 42

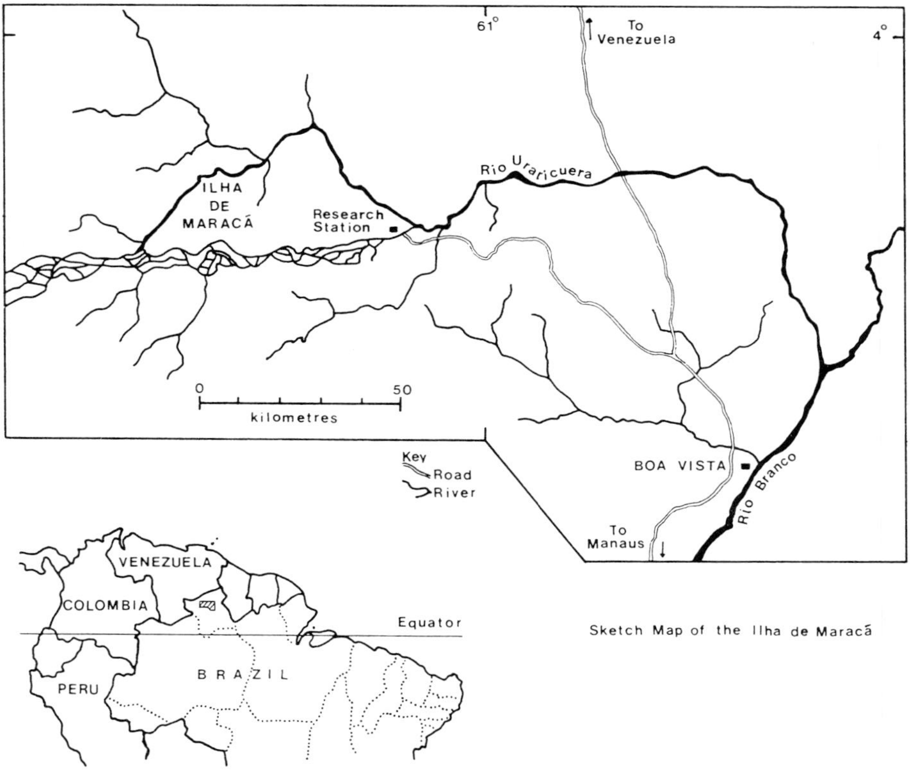

Map 1: Location of the Ilha de Maracá in Brazil and South America.

INTRODUÇÃO

Este trabalho foi elaborado como parte do Projeto Maracá 1987-88 (Instituto Nacional de Pesquisas da Amazonia, Royal Geographical Society e Secretaria Especial do Meio Ambiente). G.P. Lewis, o autor principal, é taxonomista de plantas do herbário do Real Jardim Botânico de Kew tendo trabalhado com a família Leguminosa nos últimos doze anos e desde 1981 tem coletado e estudado as leguminosas brasileiras, especialmente do Estado da Bahia, tendo tido a oportunidade de visitar o Brasil durante sete ocasiões. Foi financiado pelo Jardim Botânico de Kew (o qual designou Maracá como principal pesquisa no exterior para 1987) nas três visitas à ilha, em março-abril 1987, outubro-novembro 1987 e fevereiro 1988. P.E. Owen, o segundo autor, está atualmente cursando a Universidade de Bradford em Ciência Ambiental e trabalhou no herbário de Kew como parte de seus estudos de graduação entre agosto 1987 e agosto 1988.

O convite original para a Royal Geographical Society para trabalhar em Maracá foi feito pelo Dr Paulo Nogueira Neto, então Secretário do Meio Ambiente e diretor da Secretaria Especial do Meio Ambiente (SEMA), no final de 1983. A SEMA construiu a estação ecológica de Maracá em 1978 como parte do programa de estações ecológicas e reservas iniciado sob a lei 6.902. O convite do Dr Paulo Nogueira para que a Royal Geographical Society desenvolvesse um inventário geral sobre a vegetação, fauna, geologia, geomorfologia, solos, hidrologia e clima da ilha foi aceito e cinco programas de pesquisa foram então formulados. Os resultados apresentados neste trabalho fazem parte de um dos cinco programas, o do inventário ecológico, coordenado no Brasil pelo Dr Ângelo dos Santos do Instituto Nacional de Pesquisas da Amazonia (INPA) e na Grã Bretanha pelo Dr Jimmy Ratter do Real Jardim Botânico de Edinburgo.

História sucinta

Maracá é uma palavra de origem tupi geralmente usada para referir-se ao chocalho usado pelos índios em cerimônias religiosas, mas agora também usado para muitas formas de chocalhos incluindo aqueles mais simples que constam de uma lata cheia de pequenas pedras e servem de brinquedo para as crianças. O intrumento musical, maracas, também tem o nome derivado da mesma fonte. Anterior à conquista do Brasil pelos portugueses o termo maracá era então mais especificamente usado para aquela forma de instrumento usado pelos pagés e feiticeiros das tribos em cerimônias religiosas. Este era então esculpido toscamente na forma de uma cabeça humana com olhos, nariz, boca, orelhas e cabelos e sustentado a altura por um bastão imitando pescoço. Dentro desta cabeça eram colocadas várias ervas (provavelmente tabaco na grande maioria) e a fumaça resultante da queima deste material, e emanada pelos vários orifícios da cabeça, enebriavam os pagés como se eles tivessem bebido grandes quantidades de vinho e neste estado de embriagues eles eram capazes de fazer então suas curas e previsões para o futuro.

Certamente por muitos séculos tribos de índios viveram na Ilha de Maracá ou nas suas cercanias ao longo do rio Urariquera. Os falantes da língua Carib, Taulipángs e Saparás viveram ao norte de Maracá e os Makuxis no nordeste; os falantes da língua Aruak, Wapixanas, Maracás e Purukotós no sul e no leste e os Yanomamis e Xirianás ao alto do rio. A Ilha de Maracá foi presumivelmente assim denominada em decorrência dos índios que nela viveram.

Na década de 1780 três expedicões de famosos exploradores portuguêses subiram remando o rio Urariquera, passaram a Ilha de Maracá chegando até o rio Uraricaá e bons mapas foram produzidos destas viagens. Em 1838 o explorador britânico, alemão por nascimento, Robert Schomburgk atingiu o Urariquera no extremo norte da Ilha de Maracá e foi o primeiro a mapear o alto rio. Em 1911 o antropologista alemão Theodor Koch-Grünberg repetiu a jornada empreendida por Schomburgk subindo o rio Urariquera e chegando até a Venezuela. Ambos passaram por Maracá em dezembro e tiveram experiência semelhante e dificil em atravessar as corredeiras. Em 1925 o norte americano A. Hamilton Rice liderou uma expedição subindo o rio até a nascente e retornou passando pela ilha através do Furo de Maracá.

Os primeiros exploradores e coletores que passaram por Maracá em suas viagens não são entretanto conhecidos por terem coletado plantas na ilha, e certamente não existe nenhuma referência de coletas de leguminosas proveniente destas primeiras explorações que seja do nosso conhecimento. A primeira coleta de leguminosa em Maracá que é do nosso conhecimento foi feita em fevereiro de 1979 por W. Rodrigues, três anos após da ilha ter sido cedida à SEMA pelo governo do Território Federal de Roraíma. A coleta de Rodrigues, de número 10152, é um espécime florífero de *Fissicalyx fendleri* feita perto da estação de campo.

Grupos de estudantes liderados pelo Dr H.O. Schubart e a Dra J.M. Rankin do INPA e o Dr A. Gomes Bandeira do Museu Goeldi de Belém estiveram na ilha para curtas visitas em dezembro de 1978 e Rankin retornou e coletou plantas durante uma visita de uma semana em março de 1979. M.J.G. Hopkins e H.C. Fortune-Hopkins passaram uma semana na estação ecológica de Maracá coletando principalmente Leguminosae e estudando a associação de besouros que se alimentam das suas sementes mas nenhuma destas coletas foi vista até agora por nós. Professor A. Knob da Universidade da Amazônia junto com um grupo de estudantes de graduação fizeram, em novembro de 1980, um inventário da vegetação em duas áreas de 20 × 20 metros. Este mesmo professor retornou a ilha em dezembro de 1981 com um segundo grupo de estudantes e permaneceram cêrca de duas semanas coletando fungos e plantas superiores para o herbário de Manaus. Novamente não é do nosso conhecimento se há leguminosas incluídas nas coletas feitas pelo Prof. Knob e os grupos liderados por ele. Durante a estação chuvosa, em junho de 1986 um grupo composto pelo INPA e o Jardim Botânico de Nova York visitaram Maracá como parte do Projeto Flora Amazonica e desenvolveram um programa de coletas de plantas e inventário sob a liderança de Dr David Campbell. Alguns componentes desse grupo retornaram à ilha durante a expedição em 1987-88.

Para satisfazer as necessidades em se desenvolver o inventário ecológico de Maracá muitas picadas foram abertas na ilha e várias sub-estações foram feitas para facilitar e melhor permitir a coleta dos diversos tipos de vegetação ao longo da ilha (veja Mapa 2). Em outubro 1987, com o rio ainda cheio o suficiente para permitir a passagem de barcos através de muitos dos conjuntos de corredeiras, Mr William Milliken juntamente com Lewis e três índios Maiongonges foram capazes de circunavegar a ilha, coletando durante a jornada e parando no acampamento da cachoeira Pururamé, no extremo oeste da ilha, para coletar por dois dias nas ricas florestas de encostas sobre solo xistoso.

A família Leguminosae

A família Leguminosae (ou Fabaceae) abrange em total uns 670 gêneros e cêrca de 17.500 espécies com distribuição cosmopolita e sendo a terceira maior família de plantas superiores depois de Compositae e Orchidaceae. A variedade de hábito entre as leguminosas é muito grande podendo elas ser das maiores árvores emergentes nas florestas até ervas diminutas e efêmeras. Nas florestas de terra firme, perto da estação principal em Maracá, o *Hymenolobium petraeum* atinge 50 metros em altura e como descrito por Ducke e Black (1953) em outras áreas da Amazônia, aparece acima do dossel da mata como um buquê gigante. Isto é notado especialmente quando a árvore está em frutificação e completamente desfolhada. Ao extremo oeste da ilha, nas florestas de encosta perto da cachoeira Pururamé, *Parkia pendula* emerge sobre o dossel da mata como uma coroa plana de 40 metros de altura e, ao longo da ilha outras árvores da família Leguminosae alcançam a altura de 30 metros.

As sementes, os legumes verdes, as folhas, as raizes e as flores de muitas leguminosas fornecem alimento rico em proteina para o homem e animais em quase todas as partes do mundo. Algumas das mais finas madeiras são oriundas de espécies de leguminosas, vários jacarandás de grande valor são espécies de *Dalbergia* e conhecidos pelos trabalhos de marcenaria e entalhes, os arcos para violino de melhor qualidade ainda são feitos de *Caesalpinia echinata*, o pau brasil. Rotenone, um inseticida e veneno para os peixes é extraido de espécies de *Derris*, *Lonchocarpus* e *Tephrosia* e as raízes de *Lonchocarpus glabrescens*, uma liana comum da beira do rio em Maracá, são certamente usadas pelos índios Maiongonges que habitam as vizinhanças da ilha como "veneno" para os peixes. Alguns dos melhores copais ou resinas viscosas, largamente usadas na produção de vernizes, tintas e lacas são produzidos de árvores de leguminosas, especialmente dos generos *Hymenaea* e *Copaifera*, duas espécies de *Hymenaea* são comuns na mata de terra firme ao leste da ilha. Muitos alimentos processados são contituidos de muitas leguminosas e/ou contêm gomas ou outros aditivos extraídos de espécies de leguminosas. Os criadores de abelhas são especialmente interessados em algumas espécies de leguminosas pela qualidade do excelente mel produzido das suas flores (entre estas, *Hymenaea courbaril* e várias espécies de *Acacia*, *Albizia*, *Andira* e *Inga*, são generos ocorrentes na ilha). Legumes também produzem alguns dos importantes corantes, como por exemplo o brilhante "indigo blue" de algumas espécies de *Indigofera* e o vermelho vivo do "brasil" de *Caesalpinia echinata*. Em outras partes do mundo as leguminosas são usadas como condimentos, aromatizantes e vários outros empregos culinários; como fonte de tanino na indústria de couros, para produção de alcaçuz, perfumes e óleos e na produção de medicamentos como componentes de anticoncepcionais orais e em alguns casos como alucinógenos. *Anadenanthera peregrina*, uma das duas espécies do genero, ocorre em Maracá e é uma das plantas que tem sido largamente usada pelos índios da Amazônia na preparação de uma potente droga alucinógena. De particular significância para a moderna medicina é a recente descoberta que o componente castanospermine da leguminosa endêmica da Australia, *Castanospermum australe* está provando ser efetivo em pesquisa na batalha contra o virus da AIDS. Também de grande significância é que este composto químico castanospermine foi descoberto também no gênero *Alexa*, grupo estreitamente aparentado de *Castanospermum* e com 10 espécies endêmicas no neotrópico. *Alexa canaracunensis* é uma das árvores dominantes nas florestas do oeste da ilha e análises recentes demostraram que a castanospermine está presente não só nas sementes, mas também nas folhas, casca e na entrecasca. A combinação do fato de haver muitos indivíduos da mesma espécie muito próximos, junto com a presença do componente castanospermine como exudato resultante do ferimento da casca, pode significar que a espécie poderá ser potencialmente importante pela estração por "sangramento", semelhante a extração da borracha. A real praticabilidade deste processo obviamente precisa ser investigada pelos brasileiros.

Em direto contraste com o comumente falado, que a floresta tropical úmida é rica em espécies mas normalmente pobre em número de indivíduos de cada espécie, algumas leguminosas em Maracá ocorrem em frequência dominante cobrindo muitos hectares. *Alexa canaracunensis* juntamente com sua parente próxima *Clathrotropis macrocarpa* são dominantes na floresta de encosta da borda oeste da ilha e, espécies de *Peltogyne* chegam a ser tão dominantes (por exemplo, perto da Casa de Maracá) que a vegetação poderia ser denominada "floresta de *Peltogyne*".

Algumas espécies dos generos maiores das leguminosas são característicos de vegetações abertas e perturbadas. Elas são adaptados para o início da colonização e subsequente expansão rápida nestes habitats devido em grande partes às suas associações com bactérias fixadoras de nitrogênio ou com micorizas. Em Maracá espécies de *Vigna*, *Centrosema*, *Crotalaria*, *Zornia* e *Stylosanthes* restritos às áreas nas imediações do acampamento base e ao longo do caminho do acampamento para o rio, devem quase que certamente ter aparecido na ilha em 1978 pela inadvertida introdução pelo homem, ou distribuidas nestas áreas abertas por outros vetores naturais.

Algumas das belas e características árvores de arborização urbana nos trópicos são leguminosas — *Delonix regia*, *Caesalpinia peltophoroides* e *Caesalpinia pulcherrima* são alguns exemplos e pelo menos três delas são comuns ao longo das ruas de Boa Vista. Várias das árvores de Maracá poderiam ser igualmente exploradas como ornamentais tais como a magnífica *Cassia moschata*, a elegante *Elizabetha coccinea* e o *Macrolobium acaciifolium*.

Formato

Os gêneros e grupos taxonômicos superiores estão arranjados sistemáticamente seguindo Polhill & Raven (1981). Sob cada nome genérico são dadas referências bibliográficas para úteis revisões e monografias além de breves comentários sobre a distribuição e o número de espécies de cada gênero. Isto é seguido pela chave de identificação para as espécies. As espécies e taxa infraespecíficos estão arranjados em ordem alfabética. Abaixo de cada nome aceito como válido, estão listados os maís importantes e conhecidos sinônimos seguidos do nome vulgar, para os quais existe um índice ao final do volume. Cada planta é apresentada com uma descrição baseada em grande parte em caracteristícas diagnósticas de campo, que foram extraídas das anotações de campo e etiquêtas das coletas feitas em Maracá, ou como resultante das observações pessoais feitas por G.P. Lewis. Ocasionalmente, devido a falta de material adequado e suficiente proveniente de Maracá, outros espécimens de herbário, provenientes de outras partes da América do Sul (especialmente Guiana) foram usadas para completar detalhes essenciais, mas é necessário esclarecer que as descrições e chaves foram feitas com o uso, e portanto para a identificação, das plantas de Maracá. As descrições são seguidas de notas sobre habitat e algumas vezes também comentários sobre usos da planta, afinidades taxonômicas e problemas taxonômicos ainda não resolvidos. As citações dos espécimens observados estão colocadas a seguir. Todos os legumes conhecidos até o momento para Maracá são citados e todos os que foram vistos pelos autores estão indicados por "!". O fato do espécimen estar em flor (fl.) ou em fruto (fr.) ou ambos também está indicado. Todos os espécimens estão depositados em Kew (K) a não ser aqueles indicados pela sigla de um herbário alternativo entre parenteses. As duplicatas de material botânico, especialmente aqueles enumeradas sob a série de números de G.P. Lewis, W. Milliken e J. Ratter estão também depositadas em E, INPA, MIRR & NY. As duplicatas adicionais foram distribuidas largamente pelo Brasil, Europa e Estados Unidos da América.

Sete ilustrações em preto e branco contendo frutos de muitas das espécies estão dispersas através do texto. Também estão incluidas oito pranchas com fotografias coloridas, uma mostrando alguns dos tipos de vegetação característicos de Maracá e outras ilustrando flores e frutos de algumas espécies.

INTRODUCTION

BACKGROUND

This work was done as part of the Maracá Rainforest Project 1987–88 (Instituto Nacional de Pesquisas da Amazonia, Royal Geographical Society, Secretaria Especial do Meio Ambiente). G.P.L. is a plant taxonomist in the herbarium at Kew and has spent the last twelve years working on the family Leguminosae and since 1981 has visited Brazil seven times to collect and study these plants, mainly in the state of Bahia. He was funded by the Royal Botanic Gardens Kew (who had designated Maracá its main overseas research effort for 1987) for three visits to the island, in March–April 1987, October–November 1987 and February 1988. P.E.O. is currently at the University of Bradford studying Environmental Science and worked in the Kew herbarium as a sandwich student between August 1987 and August 1988.

The original invitation to the Royal Geographical Society to work on Maracá was sent by Dr Paulo Nogueira Neto, the then Secretary of the Environment and Director of the Secretária Especial do Meio Ambiente (SEMA), at the end of 1983. SEMA had built the Maracá ecological station in 1978 as part of a programme of research stations and reserves set up under Law 6.902. Dr Paulo Nogueira asked the Royal Geographical Society to undertake a general survey of the island's vegetation, fauna, geology, geomorphology, soils, hydrology and climate. This invitation was accepted and five research programmes were formulated with the main aim of providing detailed information about the islands complex tropical ecosystem. The results presented here fall within one of these five programmes — the ecological survey — co-ordinated in Brazil by Dr Angelo dos Santos of the Instituto Nacional de Pesquisas da Amazonia (INPA) and in the U.K. by Dr Jimmy Ratter of the Royal Botanic Garden, Edinburgh.

The project was officially launched by its Patron, HRH the Prince of Wales on 4 December 1986 at the Royal Geographical Society in London.

BRIEF HISTORY

Maracá is a Tupi Indian word generally agreed to refer to a gourd rattle used in religious ceremonies but now used for any form of rattle, including those made of tin filled with stones and used as children's toys. The musical instruments, maracas, derive their name from the same source. Prior to the conquest of Brazil by the Portuguese the term maracá was more specifically used for a special form of ceremonial instrument used by the shaman or witch doctor. This was a large gourd crudely sculpted into the form of a human head complete with eyes, ears, nose, mouth and hair and held aloft on a neck-like pole. Into this 'head' was placed various foliage (probably mostly tobacco), and the potent fumes resulting from the burning of this material emanated from the various orifices of the 'head' and inebriated the witch doctors as though they had drunk much wine. In this state of hallucination they were able to predict the future.

Certainly for many centuries tribal Indians lived on and near Maracá Island or moved along the Urariquera river. Carib-speaking Taulipáng and Sapará lived just north of Maracá, with Makuxi to the north-east; Aruak-speaking Wapixana, Maracá and Purukotó lived to the south and east and Yanomami and Xirianá lived up-river. The Ilha de Maracá was presumably named after the Indian tribe of that name and not vice-versa. Today Maracá is uninhabited but for the field station caretaker and his family, but small fazendas (farms) and mining communities are common on land either side of the island. The nearest

Indian group are the Maiongong, to the northwest of the island near the Venezuelan border.

In the 1780's three expeditions by famous Portuguese explorers paddled up the Rio Urariquera past the Ilha de Maracá and onwards to the Rio Uraricaá and all produced good maps of their travels. In 1838 the German-born British explorer Robert Schomburgk entered the Urariquera near the northern tip of Maracá and was the first to map the upper river. In 1911 the German anthropologist Theodor Koch-Grünberg repeated Schomburgk's journey up the Urariquera and on into Venezuela. They both passed Maracá in December and experienced similar difficulties with the rapids due to the low level of water in the river. In 1925 the American, A. Hamilton Rice led an expedition up the river to its source and on the return journey passed the island by the Furo de Maracá.

The early explorers and collectors who passed Maracá on their travels are not known to have collected on the island itself and certainly there are no early legume collections known to us. The first Maracá legume collection that we have seen is that of W. Rodrigues collected in February 1979, three years after the island was ceded to SEMA by the Govenor of the Federal Territory of Roraima. Rodrigues' collection, number 10152, is a flowering specimen of *Fissicalyx fendleri* from near the station base-camp.

Groups of students led by Dr H.O. Schubart and Dr J.M. Rankin of INPA and Dr A. Gomes Bandeira of the Museu Goeldi in Belém visited the station for brief periods in December 1978 and Rankin returned and made plant collections during a one week visit in March 1979. M.J.G. Hopkins and H.C. Fortune-Hopkins spent a week at the station in December 1979 collecting mainly Leguminosae and associated seed-eating beetles but we have yet to see any of these collections. Professor A. Knob, from the University of Amazonia, together with a group of graduate students, carried out a vegetation inventory of two 20 x 20 metre plots in November 1980 and returned with a second group in December 1981 to spend about two weeks collecting fungi and higher plants for the herbarium in Manaus. Again, we do not know if any legumes were included amongst those collections. During the rainy season in June 1986 a joint New York Botanical Garden — INPA team visited Maracá as part of Projeto Flora Amazonica and carried out a programme of general plant collecting and inventory work under the leadership of Dr David Campbell. Some of this group returned to the island during the 1987–88 expedition. In order to satisfy the requirement of carrying out an ecological survey of Maracá several picadas (trails) were opened up around the island and various sub-camps set up in order to permit sampling of the many different vegetation types. (See Map 2). In October 1987, with the river water still high enough to allow boat passage through most sets of rapids, Mr William Milliken together with G.P.L. and three Maiongong Indians were able to circumnavigate the island, collecting on the journey and stopping at the Pururamé waterfall camp, at the western end of the island, to collect for two days in the rich hill-slope forests overlying metamorphosed schists.

TOPOGRAPHY, GEOLOGY AND CLIMATE

The Ilha de Maracá is one of the world's largest riverine islands covering an area of some 92,000 hectares (ca. 227,000 acres or 350 square miles) or up to 100,000 hectares if some of the associated smaller islets are also included. It is approximately twice the size of the Isle of Wight being about 60 kms long by 15–25 kms wide (38 × 9½–15 miles). The island is formed by the dividing and rejoining of the Rio Urariquera (Uraricoera), a headwater of the Rio Branco, itself a tributary of the Rio Negro which meets the Rio Solimões at Manaus. At the

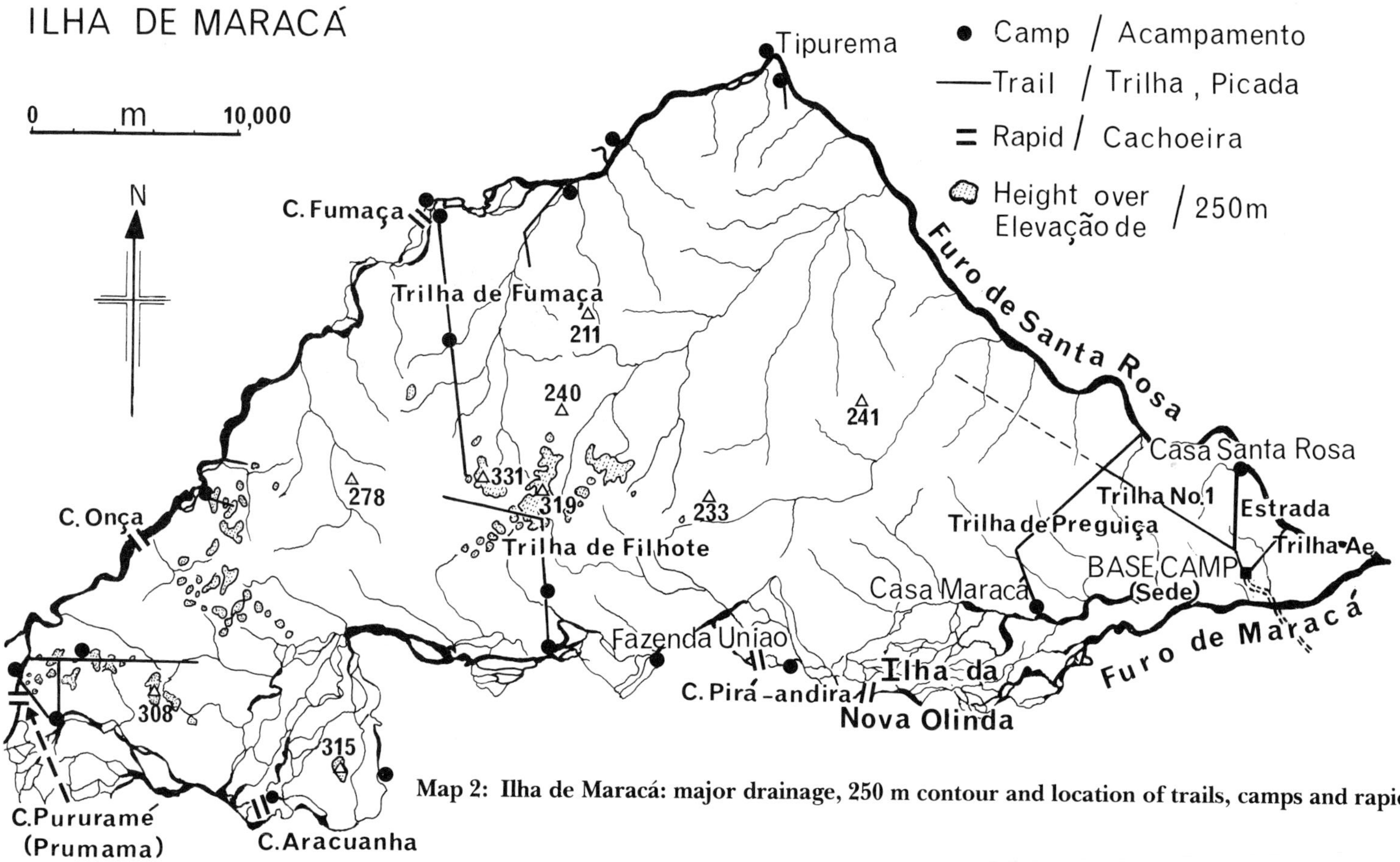

Map 2: Ilha de Maracá: major drainage, 250 m contour and location of trails, camps and rapids.

western tip of the island the Urariquera divides to become the Furo de Maracá to the south and the less treacherous Furo de Santa Rosa to the north and these rejoin at the eastern tip of the island (see maps 1 and 2). Lying between latitudes 3°15' to 3°35' north and longitudes 61°22' to 61°58' west, and approximately 130 kilometres north west of Boa Vista, the island falls within the municipality of Boa Vista in the Federal Territory of Roraima. Apart from the family manning the research station near the eastern end of the island Maracá is uninhabited.

Most of Maracá is fairly flat although in the north and west there are some low peaks reaching between 200 to just over 300 metres in altitude and to the south there are some fascinating granitic outcrops that support at least two Amazonian species of cactus (*Melocactus smithii* and *Cereus hexagonus*). The river channels are up to 50 metres wide and there are frequent, scarcely navigable rapids on each, particularly numerous on the Furo de Maracá. Even so, the eastern tip of the island is only about 30 kilometres above the limit of uninterrupted navigation from the Atlantic.

The underlying geology is of Lower Precambrian age being part of the ancient Guianan Shield, but not far to the north are the Roraima — Pacaraima sandstone escarpments that form the watershed between the Amazon and Orinoco basins. This area, known as Pan-tepuia, is characterised by flat-topped mountains known as tepuis including Mount Roraima itself. The nearest tepui to Maracá is the small and isolated Tepequem, some 30 kilometres from the northern most point of the island.

Maracá has a seasonal tropical climate with a mean annual temperature of ca. 26°C (with a range from 12 to 38°C) and an annual rainfall of between 1750 and 2250 mm. Based on RADAMBRASIL records covering a 30 year sample, about 75% of the annual rainfall at Boa Vista falls between April and August. On Maracá the wettest months are May, June and July with a dry season from November to March, but 1987 observations on precipitation supported Moskovits' findings of 1985 that some of the heaviest rainstorms occurred in the dry season. However, dry-season days are normally hot, sunny and clear and often quite windy, whereas rainy season days are generally overcast.

VEGETATION

Being uninhabited and covered by a diverse array of vegetation types and micro-habitats, Maracá is rich in flora and fauna. Most of the island is covered by undisturbed terra firme forest (humid tropical rainforest of Moskovits, 1985, and seasonal evergreen forest of G.P.L.'s field labels) but there are also areas of seasonally flooded forest (igapó), dense formations of buriti palms (*Mauritia* spp.), seasonal lakes within the forest, open wetlands, swamps and pockets of savanna (grassy campo of Moskovits, 1985, cerrado of some workers). The presence of forest and savanna and the abrupt junction between the two partly explains the rich legume flora on Maracá.

Moskovits (1985) states that the 'dense Amazonian rainforest abruptly reaches its northeastern limit on the sandy soils of Maracá, where it opens out into the bare grasslands of the "Gran Sabana". The exact origin of this savanna is not known. Some argue that it is directly related to reduced annual rainfall (Guerra 1957, Haffer 1979), others maintain that it is related to shallow, nutrient poor soils (Walter 1973)'. On the island there seems little probability that the abrupt transition from forest to savanna can be related directly to rainfall but recent research is suggesting that part of the explanation is due to a change of slope and associated height of the water table (pers. comm., various Maracá co-workers, 1987). To quote Moskovits again, 'whatever its origin, it is generally agreed that the Gran Sabana represents a natural form of xeric vegetation covering shallow soils, and that it was probably formed during the Pleistocene (RADAMBRASIL, 1975). The vegetation type is very similar to the "campos" and "cerrados" of central Brazil'.

Various collectors on Maracá have interpreted the wide range of vegetation types in different ways and in consequence have sometimes used differing terminology for the same vegetation category. After consultation with Ratter and Milliken we have chosen to standardise most of the terminology in the text of this volume. Below is provided a glossary to the main vegetation types mentioned in the text together with a cross reference to synonymous terms encountered on field labels.

BURITIZAL: a stand of palms (mainly *Mauritia* spp.); these particularly common at the eastern end of the island.

BREJO: swampland (see vazante).

CAMPINA FOREST: campina is a diminutive of campo and most usually applied to small open areas in virgin forest with white sandy soil and black humus, never flooded by river water but often with marshy areas of stagnant water resulting from heavy rains. Campina forest on Maracá is low forest with few trees over 30 cm in diameter and many multiple-trunked.

CAMPO [SEASONALLY FLOODED]: low, flat sandy areas with annual or perennial herbs and grasses, often flooded during the months of May, June and July.

CARRASCO WOODLAND: Carrasco is a vague term used in different ways by different workers but on the island refers to a vegetation similar to campina forest found along the Preguiça trail.

CAPOEIRA: ± secondary scrub resulting from the disturbance of primary vegetation by man.

RIVERINE FOREST: a general term referring to all riverside vegetation that includes an element of trees regardless of whether the area is seasonally flooded or not.

SAVANNA: open grassland on sandy soil with scattered shrubs and occasional trees, ± equivalent to cerrado of central Brazil.

TERRA FIRME FOREST: on Maracá more or less a seasonal evergreen forest of the higher slopes which are not susceptible to seasonal inundation, but the composition varies greatly over the island. On the Fumaça trail and at the southern end of the Preguiça trail (see map 2) is a tall mixed forest with rather few species, dominated by *Peltogyne gracilipes* (often to 40 m tall). This will sometimes appear as seasonally flooded forest on field labels e.g. Milliken 99–112, 116–121 and 448–452.

VÁRZEA FOREST: lowland strips of vegetation along the margins of rivers and lakes subject to seasonal flooding. There is no true várzea on Maracá and it is rarely mentioned on field labels.

VAZANTE: a scrubby tangle of vegetation always flooded seasonally and giving the appearance of abandoned allotments with herbaceous legume climbers and twiners common.

Recent counts, based mainly on material housed in the Herbarium of INPA, Manaus have shown that Amazonian Brazil contains 1255 species of legume in 152 genera (comm., M.F. da Silva et al., 1987). Of the 54 genera and 123 species of legume so far recorded on Maracá only 4 genera and about 36 species are restricted to South American equatorial rain forest and not all of these can be said to be strictly Amazonian, some showing connections with the Guianan flora. Here the reader's attention is drawn to A. Ducke and G.A. Black's excellent paper entitled 'Phytogeographical notes on the Brazilian Amazon' published in english in 1953 and in portuguese in 1954 (see bibliography for details) and including numerous observations about Amazonian Leguminosae.

THE FAMILY LEGUMINOSAE

The Leguminosae (or Fabaceae) comprising some 670 genera and about 17,500 species worldwide are the third largest flowering plant family after the Compositae and Orchidaceae. They are very variable in habit, ranging from some of the largest forest emergents to tiny ephemeral herbs. In terra firme forest, near the main station on Maracá, *Hymenolobium petraeum* attains 50 metres in height and as noted elsewhere by Ducke and Black (1953) appears above the general canopy of the forest like a giant bouquet. This show is especially accentuated in fruit when the tree is otherwise completely leafless. At the western end of the island on the hill-slope forests near the Pururamé falls *Parkia pendula* emerges above the general canopy as a flat-crowned 40 metre tree, and throughout the island other leguminous trees regularly attain a height of 30 metres.

The seeds, young green pods, leaves, roots and flowers of many legumes provide a protein rich food source for humans and animals in nearly all parts of the world and the large scale cultivation of some species produces valuable cash-crops. Some of the finest timbers come from legume species; various rosewoods from *Dalbergia* species are highly valued for cabinet work and carving and the best quality violin bows are still made from *Caesalpinia echinata* or 'pau brasil'. Rotenone, an insecticide and fish poison is extracted from species of *Derris*, *Lonchocarpus* and *Tephrosia* and the roots of *Lonchocarpus glabrescens*, a common riverside climber on Maracá, are certainly used for this purpose by the neighbouring Maiongong Indians. Some of the best copals, or viscous resins, widely used in the production of varnishes, paints and lacquers are produced from leguminous trees, particularly of the genera *Hymenaea* and *Copaifera*. Two species of *Hymenaea* are common in the terra firme forest on the eastern end of Maracá. Many processed foods contain gums extracted from various legumes and several species are prized by bee-keepers for producing some of the finest honey (amongst these, *Hymenaea courbaril* and various species of *Acacia*, *Albizia*, *Andira* and *Inga* all occur on the island). Legumes also produce some important dyes, for example, the brilliant blue 'indigo' from some *Indigofera* species and the wine red 'brasil' from *Caesalpinia echinata*. In other parts of the world legumes are used to flavour curries, pickles and other foods, as a source of tannin for the leather industry, to produce licorice, perfumes and oils, in the production of medicines, as a component of oral contraceptives and in some instances as hallucinogenic drugs. *Anadenanthera peregrina*, one of the two species in the genus, occurs on Maracá and this is one of the plants that has been widely used by the Amazonian Indians in the preparation of a potent hallucinogenic snuff. Of particular significance to modern medicine is the recent discovery that the compound castanospermine, from the Australian endemic legume *Castanospermum australe*, might prove effective in the battle against the AIDS virus. Of equal significance is that this plant chemical has now been discovered in the closely related genus *Alexa* with 10 species endemic to the neotropics. *Alexa canaracunensis* is one of the dominant tree species in the forests at the western end of the island and recent analyses of material collected on Maracá have shown that castanospermine is present not only in the seeds but in leaves, bark and sap. The combination of having several individuals of one species in close proximity, together with the presence of castanospermine in exudate resulting from a bark slash means that this species is potentially well suited to tapping, similar to the collection of rubber. The real practicalities of this obviously need further investigation by Brazilians.

In direct contrast to the common understanding that all tropical rain forest is species rich but usually poor in numbers of individuals of any one species, some legumes on Maracá occur in dominant stands covering several hectares. *Alexa canaracunensis* together with the closely related *Clathrotropis macrocarpa* are dominant on the hill-slope forests in the west and species of *Peltogyne* are so dominant in some areas (e.g. near the Casa de Maracá) that the vegetation can truely be termed *Peltogyne* forest.

Many species in the larger legume genera are highly characteristic of open and disturbed sites. They are well suited to early colonization of and subsequent rapid expansion in such habitats due, in a large part, to their associations with either nitrogen fixing bacteria or ectomycorrhiza. On Maracá, species of *Vigna*, *Centrosema*, *Crotalaria*, *Zornia* and *Stylosanthes* restricted to sites around base-camp or along the track from the camp to the river have almost certainly appeared on the island since 1978 either introduced inadvertently by man or distributed to these open sites by other natural vectors.

Many of the beautiful and characteristic street trees of the tropics are legumes — *Delonix regia*, *Caesalpinia peltophoroides* and *Caesalpinia pulcherrima* to name but three that are common along the streets of Boa Vista. Several of the trees on Maracá could equally be used as ornamentals e.g. the magnificently showy *Cassia moschata* and the elegant *Elizabetha coccinea* and *Macrolobium acaciifolium*.

FORMAT

The genera and higher taxonomic groups are arranged systematically following Polhill and Raven (1981). Beneath each generic name are given bibliographic references to useful revisions and monographs and brief notes about the distribution and size of the genus worldwide. These are followed by a key to species. The species and infraspecific taxa are arranged alphabetically. Beneath each accepted name are listed commonly used and nomenclaturally important synonyms followed by known vernacular names for which there is an index at the end of the volume. Each plant is provided with a description based largely on diagnostic field characters either extracted from Maracá field labels or from personal observations by G.P.L. Occasionally, due to a lack of adequate material from Maracá, herbarium specimens from other parts of South America (usually British Guiana) have been used to fill in essential details but it should be stressed that both the descriptions and keys are designed essentially to apply to Maracá collections only. The description is followed by notes on habitat and sometimes also notes on plant uses, taxonomic affinities or unresolved taxonomic problems. Specimen citations follow the notes. All known Maracá legumes are cited and all have been seen by the authors as indicated by a '!'. Whether the specimen is in flower (fl.) or fruit (fr.) or both is also recorded. All specimens are housed at Kew (K) unless otherwise indicated by an alternative herbarium acronym in parentheses. Duplicates of most collections, especially those of G.P.L., Milliken and Ratter are also housed at E, INPA, MIRR and NY. Additional duplicates have been widely distributed throughout Brazil, Europe and the U.S.A.

Seven black and white figures illustrating the fruits of a large percentage of the species are interspersed throughout the text. There are eight colour plates, one showing some of the habitat types and the others illustrating flowers and fruits of some of the species.

SYNOPSIS OF TRIBES

1	CASSIEAE	:	genera 1, *Dialium* to 4, *Chamaecrista*
2	CERCIDEAE	:	genus 5, *Bauhinia*
3	DETARIEAE	:	genera 6, *Peltogyne* to 8, *Elizabetha*
4	AMHERSTIEAE	:	genus 9, *Macrolobium*
5	PARKIEAE	:	genus 10, *Pentaclethra* and 11, *Parkia*
6	MIMOSEAE	:	genera 12, *Entada* to 14, *Mimosa*
7	ACACIEAE	:	genus 15, *Acacia*
8	INGEAE	:	genera 16, *Inga* to 20, *Zygia*
9	SWARTZIEAE	:	genera 21, *Swartzia* and 22, *Zollernia*
10	SOPHOREAE	:	genera 23, *Acosmium* to 26, *Clathrotropis*
11	DALBERGIEAE	:	genera 27, *Andira* to 32, *Fissicalyx*
12	ABREAE	:	genus 33, *Abrus*
13	MILLETTIEAE	:	genus 34, *Lonchocarpus* and 35, *Tephrosia*
14	ROBINIEAE	:	genus 36, *Coursetia* and 37, *Sesbania*
15	INDIGOFEREAE	:	genus 38, *Indigofera*
16	DESMODIEAE	:	genus 39, *Desmodium*
17	PHASEOLEAE	:	genera 40, *Mucuna* to 49, *Eriosema*
18	AESCHYNOMENEAE	:	genera 50, *Aeschynomene* to 53, *Stylosanthes*
19	CROTALARIEAE	:	genus 54, *Crotalaria*

KEY TO GENERA

The key is divided into five groups based on foliage characteristics:

Leaves unifoliolate (sometimes bilobed to the base)	**Group 1**
Leaves bifoliolate	**Group 2**
Leaves trifoliolate	**Group 3**
Leaves pinnate	**Group 4**
Leaves bipinnate	**Group 5**

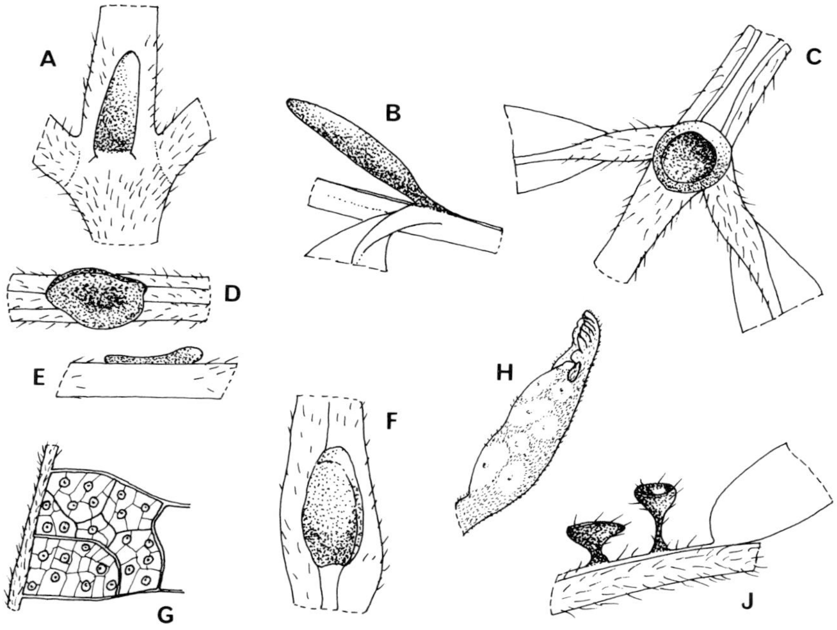

Gland types referred to in the key and text

A & **B**: peg-shaped between leaflets, **C–F**: discoid or cup-shaped between leaflets or on petiole, **G**: under-surface of leaflet, **H**: on bracts, **J**: pinhead-shaped on petiole.

A: *Senna macrophylla* var. *gigantifolia* × 4, **B**: *Senna obtusifolia* × 14, **C**: *Inga alba* × 6, **D** & **E**: *Acacia paniculata* × 8, **F**: *Parkia pendula* × 8, **G**: *Centrolobium paraense* × 6, **H**: *Alexa canaracunensis* × 4, **J**: *Chamaecrista nictitans* var. *disadena* × 8.

Group 1.

(Leaves unifoliolate)

1. Herbs or shrublets to ca. 2 m of campo or savanna; leaflets linear-oblong to lanceolate, neither bilobed nor with a dentate margin; flowers papilionaceous, zygomorphic; most stamens of each flower fused into a staminal sheath:
  2. Stipules lanceolate, papery, persistent but never decurrent; lower face of leaflets with yellow gland dots (just visible with a × 10 lens); calyx not bilabiate; keel petals not strongly beaked; fruit ± compressed, the valves pubescent with long spreading hairs (Fig. 7D) ... ... **49. Eriosema simplicifolium**
  2. Stipules characteristically enlarged, foliaceous, decurrent (the stems thus winged between the nodes) or stipules lacking; lower face of leaflets without gland dots; calyx ± bilabiate; keel petals strongly beaked; fruit inflated, glabrous or almost so (e.g. Plate 8F) ... ... **54. Crotalaria**
1. Woody climbers, shrubs or trees of riverine and terra firme forest; leaflets either bilobed (sometimes to the base) or narrowly elliptic to oblanceolate with a dentate margin; flowers not papilionaceous, ± actinomorphic; stamens free:
  3. Leaves bilobed (sometimes to the base) their margins entire; secondary venation palmate; fruit dehiscent, more than 2-seeded; climbers or shrubs ... ... **5. Bauhinia**
  3. Leaves not bilobed, their margins dentate by prickle-like extensions of the marginal vein; secondary venation never palmate; fruit drupaceous, indehiscent, 1–2-seeded; trees ... ... **22. Zollernia**

Group 2.

(Leaves bifoliolate)

1. Trees, rarely shrubs, 3–40 m tall; stipules early deciduous; leaflets lacking gland dots and not several-costate from base; inflorescence an axillary or terminal panicle:
  2. Fruit indehiscent, ovoid to cylindrical, woody, 1–several-seeded; petiolules twisted (e.g. Fig. 1N) ... ... **7. Hymenaea**
  2. Fruit 2-valved, dehiscent, laterally compressed, sublunate to subtriangular, coriaceous, 1(–2)-seeded; petiolules not or scarcely twisted (e.g. Fig. 1C) ... ... **6. Peltogyne**
1. Herbs or small shrublets, up to 80 cm tall; stipules persistent, papery; leaflets either gland-dotted or several-costate from the base; flowers axillary, either solitary or in spikes:
  3. Stipules cordate at base but not appendaged; leaflets several-costate, without glands; flowers long-pedicellate, solitary in axils; fruit dehiscent, the valves twisting (e.g. Fig. 1M) ... ... **4. Chamaecrista**
  3. Stipules strongly appendaged below their point of attachment; leaflets with scattered, punctate glands on the lower face, 1-costate; flowers concealed by foliaceous bracts in axillary spikes; fruit a loment of 4–6 bristly articles (Fig. 6T) ... ... **52. Zornia**

Group 3.

(Leaves trifoliolate)

1. Leaves digitately trifoliolate (i.e leaflets all arising from the same point, rachis lacking):
  2. Leaflets under 20 × 5 mm; flowers and 1–2-seeded fruits (Fig. 6S) obscured by bristly bracts; plant prostrate ... ... **53. Stylosanthes**
  2. Leaflets well over 25 × 5 mm; flowers and many-seeded fruits not obscured by bristly bracts:
    3. Leaves subsessile, the petiole under 5 mm long; secondary venation of leaflets closely parallel; leaflets strongly discolourous, the lower face densely appressed grey pubescent; flowers single or paired in leaf axils; fruits linear-oblong, laterally compressed, densely pubescent with grey and brown hairs (Fig. 6D) ... ... **35. Tephrosia**
    3. Leaves petiolate, the petiole usually well over 2 cm long; secondary venation of leaflets not parallel; leaflets not strongly discolourous, the lower face finely appressed-pubescent; many-flowered racemes terminating in a cluster of curled bracts; fruit inflated, glabrescent (Fig. 6F) ... ... **54. Crotalaria micans**
1. Leaves pinnately trifoliolate (i.e. the terminal leaflet arising above the other two, rachis present):
  4. Herbs or shrubs with prostrate or erect stems:
    5. Flowers pink, mauve, purple or white; leaves lacking gland dots and glandular hairs:
      6. Flowers purple; wings expanded, longer than the standard; fruit a long, cylindrical, dehiscent pod (Fig. 6U) ... ... **47. Macroptilium**
      6. Flowers pink, mauve or white; standard longer than the unexpanded wings; fruit a loment (Fig. 6H–M) ... ... **39. Desmodium**
    5. Flowers yellow; leaves with gland dots or glandular hairs:
      7. Leaflets under 20 × 5 mm, lacking gland dots; stipels lacking; prostrate herb; inflorescence a few-flowered head; flowers and 1–2-articled loments (Fig. 6R & S) obscured by bristly bracts ... ... **53. Stylosanthes**
      7. Leaflets well over 30 × 5 mm, gland-dotted on lower face; stipels present but sometimes early caducous; erect herbs or shrublets; inflorescence a many-flowered raceme; flowers and fruits not obscured by bracts:
        8. Leaves subsessile; leaflets more than 5 × as long as wide; calyx lobes subequal, shorter than tube; flowers under 10 mm long; fruits pubescent with long, fine hairs, glandular hairs lacking ... ... **49. Eriosema violaceum**
        8. Leaves petiolate, the petiole several times longer than the rachis; leaflets under 3 × as long as wide; calyx lobes unequal, the lowest longer, all longer than tube; flowers over 10 mm long; fruits pubescent with short hairs and longer glandular hairs intermixed (Fig. 7C) ... ... **48. Rhynchosia**
  4. Climbers with twining stems:
    9. Flowers under 10 mm long, not resupinate, blue (the standard sometimes with a central yellow blotch); fruit distinctly but shallowly transversely grooved (the grooves corresponding to strong internal septa), under 10 mm wide (Fig. 7E & F) ... ... **43. Calopogonium**
    9. Flowers over 10 mm long, white, pink, mauve, purple, magenta or yellow, if blue then resupinate; fruit not transversely grooved or if so then over 10 mm wide:

10. Flowers resupinate, the standard thus acting as an insect landing-platform; fruits with at least one ridge running lengthwise along the valves:
  11. Calyx strongly 2-lipped, the upper two lobes much larger than the lower 3; fruit over 10 cm long, each valve with one ridge parallel to and ca. 5 mm from the upper margin (Fig. 7L) ... ... **42. Canavalia**
  11. Calyx not strongly 2-lipped, although the teeth sometimes very unequal in length; valves of fruit each with two ridges or if only one then this ± central and fruit ca. 5 cm long:
    12. Flowers predominantly white; standard lacking a spur; fruit valves with one ridge only (Fig. 7M); leaflets generally rounded to emarginate at apex ... ... **45. Clitoria**
    12. Flowers predominantly magenta, pink, blue or buff-coloured; standard conspicuously spurred on back; fruit valves with 2 ridges (Fig. 7G–J); leaflets generally acute at apex ... ... **44. Centrosema**
10. Flowers not resupinate; fruit valves without longitudinal ridges although the upper margin sometimes thickened:
  13. Inflorescence pendulous on peduncles up to 75 cm long, persistent woody nodes alternating along a ziz-zag thickened rachis; standard shorter than keel petals; fruit with numerous undulating transverse ridges and stinging hairs (Fig. 7A); whole plant drying blackish ... ... **40. Mucuna**
  13. Inflorescence ± erect, the peduncle much shorter, the rachis not zig-zag; standard equalling or longer than keel; fruit lacking transverse ridges and stinging hairs; at least the foliage drying greenish or brownish:
    14. Robust woody lianas; flowers predominantly lilac, pinkish or purplish (the standard usually with a central yellow blotch); upper 2 calyx teeth fused to form a lobe so that calyx appears ± 4-toothed; fruit woody with the upper margin thickened (Fig. 7B & K); leaflets not lobed ... ... **41. Dioclea**
    14. Twining herbs; flowers yellow or white (Plate 7B & F) with a central yellow blotch; calyx 5-toothed; fruit chartaceous to subcoriaceous (Fig. 5L–N & Plate 7C & D); leaflets ± lobed at base ... ... **46. Vigna**

Group 4.

(Leaves pinnate)

1. Herbs, either erect, prostrate or twining, if somewhat shrubby then erect, aquatic and eglandular with leaflets under 20 × 5 mm and flowers papilionoid:
  2. Creeping or twining herbs:
    3. Prostrate herb (Plate 8C) arising from a central tap root; leaflet blades asymmetric about the midvein; leaflet margins dentate; standard yellow becoming brown and papery and persisting as an 'envelope' around the plicate fruit which breaks up into articles (Fig. 6G) ... ... **51. Soemmeringia**
    3. Twiner; leaflet blades symmetric about the midvein; leaflet margins entire; corolla white, pink or purplish; standard not persisting; fruit a dehiscent pod (Fig. 6E) ... ... **33. Abrus**
  2. Erect herbs; if somewhat shrubby (to ca. 2.5 m) then aquatic and pithy and lacking foliage glands:

4. Leaf rachis and petiole eglandular; flowers papilionoid; most stamens fused into a staminal sheath:
   5. Mature leaves 5–7-foliolate; leaflets oblanceolate with a pubescence of medifixed hairs (Fig. 6Cii) corolla pinkish red (Plate 8E); staminal sheath persistent; fruit a dehiscent pod (Fig. 6Ci) ... ... **38. Indigofera**
   5. Mature leaves over 16-foliolate; leaflets linear-lanceolate to oblong, lacking medifixed hairs; corolla predominantly yellow; if staminal sheath persistent then fruit a loment:
      6. Stipels absent; corolla up to 10 mm long; fruit a loment (Fig. 6 N–Q) ... ... **50. Aeschynomene**
      6. Stipels present; corolla over 25 mm long; fruit an elongate dehiscent pod (Fig. 6A) ... ... **37. Sesbania**
4. Leaf rachis with peg-shaped glands between at least the basal pair of leaflets; flowers caesalpinioid; stamens free ... ... **3. Senna obtusifolia**

1. Trees, robust shrubs or woody climbers; if small shrubs to ca. 3 m only then leaflets well over 40 × 20 mm and flowers caesalpinioid:
   7. Leaflets with numerous orange, scale-like glands on lower face; bracts and bracteoles densely felty tomentose with orange glands intermixed; fruit a woody samara with a spiny seed cavity and a persistent stylar spine arising from the wing just above the seed chamber (Fig. 5G) **31. Centrolobium**
   7. Leaflets, bracts and bracteoles without gland dots; fruit not as above:
      8. Leaflets alternate:
         9. Flowers apetalous; fruits globose to ellipsoid, 1–2-seeded, indehiscent; tree with glabrous leaflets ... ... **1. Dialium**
         9. Flowers with petals; fruits various but not as above; if leaflets glabrous then plant a woody climber:
            10. Petals yellow veined red; stamens dimorphic, the lower 3 with S-shaped filaments; fruits indehiscent, cylindrical, sausage-shaped, up to 50 × 1.5 cm, up to 100-seeded (Fig. 1J); leaflets asymmetric about the midvein at the truncate base; leaves lacking glands ... ... **2. Cassia**
            10. Petals white, cream, pink or lilac (the standard sometimes with a yellow blotch); stamens homomorphic; if fruits indehiscent then 1–2-seeded; leaflets symmetric at the cuneate to rounded or cordate base:
                11. Stipels present; keel petals rostrate; fruit dehiscent, glabrous, 14–18-seeded (Fig. 6B) ... ... **36. Coursetia**
                11. Stipels absent; keel not rostrate; if fruit dehiscent then up to 6-seeded:
                    12. Flowers at anthesis 6–8 cm long, pseudocaesalpinioid; stamens free; petals creamish; bracts fleshy, pear-shaped, glandular; fruits dehiscent, 2–5(–6)-seeded; seeds embedded in a white pithy endocarp (Fig. 4H, Plate 5C & D); tree; leaflets with acuminate drip-tips; small, obscure glands sometimes present between the leaflets ... ... **25. Alexa**
                    12. Flowers under 2 cm long, papilionoid; most stamens fused into a staminal sheath; bracts neither fleshy nor glandular; fruits indehiscent, 1(–2)-seeded; if leaflets acuminate then plant a climber; glands lacking on leaves:
                        13. Standard petal glabrous; fruit velvety brown, 1-seeded, orbicular to subreniform, indehiscent (Fig. 5F); leaflets tomentose on lower face, drying brownish ... ... **30. Dalbergia**
                        13. Standard petal hairy on outer face; fruit a samara (Fig. 5B, C & E) or if ± as above then leaflets glabrous, drying blackish ... ... **29. Machaerium**
      8. Leaflets opposite to slightly subopposite:

14. Petals valvate in bud, fused for at least ½ their length into a tubular corolla; stamens numerous, homomorphic, fused for part of their length into a staminal tube; discoid or cup-shaped glands present between all leaflet pairs ... ... **16. Inga**
14. Petals imbricate in bud or flowers with a single petal only, petals essentially free or the two lowest petals partially fused; if stamens well over 10 in number then usually dimorphic and not fused; if foliage glands present then these peg or pin-shaped or sunken in obscure cushion-like swellings below the petiolules:
  15. Flowers with a single (rarely 2) petal(s); leaf rachis usually partially winged and/or flower buds globose, splitting irregularly at anthesis and/or seeds arillate:
    16. Petals white; stamens 3, homomorphic; mature leaves generally with 20 or more leaflets; fruit laterally compressed, woody, orbicular, tardily dehiscent, glabrous, drying orangish brown (Fig. 1B, Plate 2D); seeds flat, lacking an aril ... ... **9. Macrolobium**
    16. Petal(s) yellow; stamens (10–) many, usually dimorphic; mature leaves with 1–13 leaflets; fruits without the above combination of characters; seeds arillate (Fig. 4B & C, Plate 5A & B) ... ... **21. Swartzia**
  15. Flowers with 5 petals; leaf rachis lacking wings; seeds without an aril:
    17. Flowers white, cream or yellow:
      18. Leaves with peg-shaped glands between at least the lower leaflet pair or with pinhead-shaped glands on the petiole or small, sunken glands on cushion-like structures below the petiolules (these often obscure and appearing as small black dots on dried material; use a × 10 lens):
        19. Petiole supporting 1–2 pinhead-shaped, stipitate glands; inflorescence few-flowered, supra-axillary, much shorter than leaf that subtends it; leaves multifoliolate; leaflets under 20 × 5 mm; shrub ... ... **4. Chamaecrista**
        19. Petiole lacking stipitate, pinhead-shaped glands; inflorescence a showy, terminal raceme or panicle, if axillary then many-flowered and leaves with only 2 pairs of leaflets and these over 10 × 5 cm:
          20. Leaves with sunken glands between some leaflets; upper stems hollow, housing ants; flowers papilionoid in bud, pseudo-caesalpinioid at anthesis; bracts fleshy, covered in numerous small circular glands; stamens homomorphic; calyx tubular, truncate; hypanthium present; petals cream (drying purplish); fruit dehiscent, velvety, up to 5-seeded (Fig. 4H, Plate 5C & D) ... ... **25. Alexa**
          20. Leaves with peg-shaped glands between at least the lowest leaflet pair; stems not hollow; flowers caesalpinioid; bracts lacking glands; stamens dimorphic; calyx with ± free petaloid sepals; hypanthium lacking; petals yellow (drying yellow or creamish); fruit lacking a velvety tomentum, over 5-seeded ... .. **3. Senna**
      18. Leaves eglandular:
        21. Stamens mostly fused into a staminal sheath:
          22. Leaves with stipels; calyces campanulate ... ... **34. Lonchocarpus sp. aff. negrensis**
          22. Leaves without stipels; calyces spathulate; each valve of fruit with a wide papery wing (Fig. 5J) ... ... **32. Fissicalyx**
        21. Stamens mostly free; leaves without stipels:
          23. Stamens dimorphic; fruits over 5-seeded ... ... **3. Senna**

23. Stamens homomorphic; fruits 1–3-seeded:
24. Leaflets rounded to retuse at apex; flowers ± pseudocaesalpinioid; fruit ± indehiscent (Fig. 4A & G); seeds orangish brown, subreniform … … … … … … … … … … … **23. Acosmium**
24. Leaflets acute to acuminate at apex; flowers papilionoid; fruits dehiscent; seeds black, globose … … … … **24. Ormosia flava**
17. Flowers pink, purple, bluish or scarlet (although standard sometimes with a central yellow, white or greenish blotch):
25. Woody lianas; inflorescences nodose; fruit indehiscent, chartaceous, 1–2-seeded and lacking an obvious marginal wing (e.g. Fig. 6V & W), or woody, several-seeded and sublomentaceous, ± breaking up into 1-seeded segments (Fig. 5A, Plate 6E) … … … **34. Lonchocarpus**
25. Trees or large shrubs; inflorescences not nodose; if fruit chartaceous then one or both margins winged:
26. Fruit dehiscent:
27. Seeds scarlet or bicoloured scarlet and black (e.g. Plate 6A), globose, under 15 mm long, the testa hard … … **24. Ormosia**
27. Seeds brown or orangish, discoid or subreniform, over 20 mm long, the testa thin and papery:
28. Flowers 10–15 mm long, papilionoid, petals purplish, hypanthium lacking … … … … … … … … … **26. Clathrotropis**
28. Flowers over 30 mm long, caesalpinioid, petals scarlet, hypanthium massive (Plate 2E) … … … … … … **8. Elizabetha**
26. Fruits indehiscent:
29. Fruit drupaceous, the mesocarp breaking down to form a malodorous pulp (e.g. Fig. 5H) … … … … … … … … **27. Andira**
29. Fruits chartaceous or subcoriaceous, laterally compressed, lacking a pulpy mesocarp:
30. Fruit purplish, glabrous, finely reticulately veined, twisted at the base above the stipe, both margins winged (Fig. 5K, Plate 6C) … … … … … … … … … … … … … … … … **28. Hymenolobium**
30. Fruit greyish or brownish, tomentose, venataion obscure, lacking a basal twist, at most only one margin winged … … … … … … … … … … … … … … … … … **34. Lonchocarpus**

Group 5.

(Leaves bipinnate)

1. All leaves with only one pair of pinnae:
2. Shrubs to medium-sized trees, lacking prickles; fruit never breaking up into 1-seeded articles; stamens more than 10 per flower … … … … … **20. Zygia**
2. Shrubs or climbers with recurved prickles; fruit breaking up into 1-seeded articles (Fig. 2F–J); flowers with 10 or less stamens … … … … **14. Mimosa**
1. Most leaves with 2 or more pairs of pinnae:
3. Robust woody liana without prickles; inflorescence a very large brush-head-like panicle of spikes (Plate 3F); fruit ca. 25–30 cm long, breaking up into 1-seeded, papery, winged, endocarp 'packages' to leave a persistent replum (Fig. 2B) … … … … … … … … … … … … … … … … … … **12 Entada**
3. Shrubs to large trees, if a climber then armed; if inflorescence a panicle then made up of heads; fruit much shorter, if breaking up then into unwinged exocarp 'packages':
4. Young branchlets armed with recurved scattered prickles; climbers, shrubs or small trees:

5. Foliage gland present on petiole; stamens well over 10 per flower; if fruit with persistent replum this without prickles or bristles ... **15. Acacia**
5. Foliage glands absent; stamens 10 or less per flower; fruit breaking up to leave a persistent, prickly or bristly replum (e.g. Fig. 2F) ... **14. Mimosa**
4. Young branchlets unarmed, although if a tree the trunk sometimes with spine-tipped warty protuberances; medium to large-sized trees, rarely scandent or vine-like:
6. Foliage glands lacking; flowers in axillary spikes; fruit an explosively dehiscent woody pod with valves coiling up after dehiscence (Fig. 2A) ... **10. Pentaclethra**
6. Foliage glands present at least on petiole or between lowest pair of pinnae; flowers in capitate or corymbose heads; fruits various but not explosively dehiscent:
7. Stamens of fertile flowers 10 per flower; peduncle with an involucre of bracts or pendulous:
8. Inflorescence pendulous on peduncles up to 1 m long; flowers dimorphic; calyx lobes imbricate in bud; fruit with sticky, pulpy mesocarp; seeds discoid, unwinged; trunk lacking warty protuberances ... **11. Parkia**
8. Inflorescence made up of fascicles of heads; peduncles under 5 cm long, each bearing an involucre of bracts; flowers homomorphic; calyx lobes valvate in bud; fruits lacking pulpy mesocarp; seeds flat, round, narrowly winged; trunks of young trees with warty protuberances ... **13. Anadenanthera**
7. Stamens more than 10 per flower; peduncle neither pendulous nor with an involucre of bracts:
9. Pinnae in 2–5 pairs per leaf:
10. Fruits breaking up into 1-seeded articles (e.g. Plate 4D) ... **18. Albizia**
10. Fruits dehiscent along one margin, not breaking up ... **17. Abarema**
9. Pinnae in 6–many pairs per leaf:
11. Fruits ± straight, many times longer than wide, the valves plano-compressed or undulate, dehiscent or breaking up into 1-seeded segments; seeds uniseriate ... **18. Albizia**
11. Fruits coiled, woody, indehiscent (Fig. 3A & B); seeds biserially arranged ... **19. Enterolobium**

1. **DIALIUM** L.

Rojo's recently prepared monograph awaits publication.
About 40 spp., pantropical, mostly W. Africa.

Dialium guianense (Aubl.) Sandw. in Lloydia 2(3): 184 (1939).
Arouna guianensis Aublet, Pl. Guian. 1: 16 (1775); *Dialium divaricatum* Vahl (1805).
"Jutaí pororoca."

Tree to 15 m (–30 or even 40 m fide 'Fruits of the Guianan Flora'); bark greyish or reddish brown; stipules deciduous; leaves pinnate, (4–)5–7-foliolate; leaflets alternate, glabrous, chartaceous to coriaceous, ovate to lanceolate, accuminate;

Plate 1.
A Brejo near base-camp
B Igarapé on north of island
C Terra firme forest near base-camp
D Riverine forest at west end of island
E Riverine forest near Pururamé falls
F As for E

Plate 2. **A** Cassia moschata **B** Bauhinia ungulata

C Bauhinia glabra **D** Macrolobium acaciifolium

E Elizabetha coccinea var. oxyphylla **F** Fruits of E

Plate 3. **A** *Hymenaea parvifolia* **B** *Hymenaea* sp. — seedlings

C *Peltogyne paniculata* subsp. *pubescens* **D** *Peltogyne gracilipes*

E *Anadenanthera peregrina* var. *peregrina* **F** *Entada polystachya*

Plate 4. **A** Albizia polyantha

B Albizia polyantha — one of the fruit forms

C Zygia glomerata

D Albizia corymbosa

E Inga sp. aff. ingoides

F Inga splendens

Plate 5. **A** Swartzia grandifolia — **B** Swartzia laurifolia

C Alexa canaracunensis — **D** Dehisced fruits of C showing developing and ripe seeds

E Clathrotropis macrocarpa — **F** As for E

Plate 6. **A** Ormosia smithii **B** Hymenolobium petraeum

C Fruits of B **D** Lonchocarpus glabrescens

E Fruits of D **F** Machaerium inundatum

Plate 7. **A** Andira surinamensis

B Vigna lasiocarpa

C Fruits of B

D Vigna adenantha

E Eriosema simplicifolium var. simplicifolium

F Flowers of D

Plate 8. **A** Canavalia grandiflora **B** Centrosema plumieri
C Soemmeringia semperflorens **D** Coursetia ferruginea
E Indigofera lespedezioides **F** Crotalaria pilosa

inflorescence a terminal panicle; buds brown tomentose; flowers small, greenish, apetalous; anthers yellow, the calyx lobes usually reflexing; fruit ellipsoid or globose, dark green to brown, ± indehiscent, glabrous, the exocarp fragile, mesocarp pulpy, 1(–2)-seeded.

Terra firme forest; riverine forest.

The pulpy mesocarp is edible; in the Guianas spider monkeys eat the fruits and disperse the seeds.

Near to Cachoeira de Fumaça, Furo de Santa Rosa, *Milliken et al.* 717 (fl. & fr.)!; Proctor Plot 2 [near base-camp], tree 0067, *Ratter et al.* 5611V (sterile)!

2. **CASSIA** L.

Irwin & Barneby in Mem. New York Bot. Gard. 35(1): 4–63 (1982).
15 species native or spontaneous in the Americas.

Cassia moschata Kunth, Nov. Gen. & Sp. 6: 338 (1824). Fig. 1J, Plate 2A.
"Acacha", "Mari-mari".

Tree to 12 m, DBH ca. 15–60 cm, branching of 12 m tree commencing 1 m from ground; bark grey, smooth, with horizontally arranged lenticels (dots and dashes), sometimes peeling in slabs; slash pink or reddish; young foliage softly tomentose; stipules semisagittately 2-lobed, early deciduous; leaves pinnate; leaflets alternate. discolourous, upper face shiny, glabrescent, lower face paler, softly pubescent; inflorescence a pendant raceme; calyces deep reddish purple, sepals reflexed at anthesis; petals buttercup-yellow, veined red, ageing orangish red; stamens 10, dimorphic, the longer three with larger anthers and deep yellow ± S-shaped filaments, the 4 central stamens yellow and the 3 antherodes white; fruit indehiscent, cylindrical, sausage-shaped, up to 50 × 1.5 cm, thinly pilosulose, maturing black, up to ± 100-seeded, the seeds in septate, pulpy chambers.

Trackside scrub (perhaps planted or as a remnant of vegetation clearance); terra firme forest; seasonally flooded várzea forest.

A magnificently elegant tree with two splendid examples in full flower around base-camp in March 1987 but undoubtedly native to Maracá and seen occasionally in forest bordering the Furo de Maracá and Furo de Santa Rosa at the eastern end of the island. Several species of insect, including bees in the genera *Xylocopa* and *Bombus*, visit the open flowers. Not seen in Boa Vista but certainly worthy of cultivation as an ornamental street tree.

Estrada from base-camp to Rio Urariquera, *Lewis* 1423 (fl. & fr.)!; near Casa de Maracá, *Milliken et al.* 110V (sterile)!; between base-camp and the 'ferry', *Ratter et al.* 5498 (fl.)!; Proctor plot 5, tree 0465, *Ratter et al.* 5678V (sterile)!

3. **SENNA** Mill.

Irwin & Barneby in Mem. New York Bot. Gard. 35 (1 & 2): 64–635 (1982).
202 species in the neotropics.

1. Leaves with 2–3 pairs of leaflets:
  2. Herbs; inflorescence 1–2-flowered in axils of upper leaves; foliage gland peg-like; leaflets obtuse apically ... ... **S. obtusifolia**
  2. Shrubs; inflorescence a many-flowered axillary raceme or terminal panicle; leaflets acute apically:

3. Terminal leaflets up to ca. 15 cm long; inflorescences axillary racemes compounded into an expansive terminal pseudopanicle: flowers large and showy, up to 7 cm across; petal blades finely puberulous on both faces ... ... **S. bacillaris** var. **bacillaris**
3. Terminal leaflets well over 16 cm long; inflorescences axillary racemes, much shorter than the leaves that subtend them; flowers up to 4 cm across; petals glabrous but for a few hairs on the veins of the outer face ... ... **S. macrophylla** var. **gigantifolia**

1. Leaves with 5 or more pairs of leaflets:
4. Median leaflets up to 25 × 7 mm (usually smaller), strongly discolourous on drying; a conical gland between the basal leaflet pair; both faces of leaflet blade glabrous or almost so ... ... **S. multijuga** var. **multijuga**
4. Median leaflets over 30 × 15 mm (usually much larger); petiolar gland lacking; at least lower face of leaflet blade finely appressed pubescent:
5. Expanded flowers subcorymbose on long pedicels, 20–30(–40) mm long; fertile stamens 7; leaflets acute, nitid on upper face ... ... **S. silvestris** var. **silvestris**
5. Expanded flowers in congested racemes, these often aggregated into candelabriform panicles, the pedicels up to 11 mm long; fertile stamens 2; leaflets obtuse to rounded or emarginate at apex, dull on upper face:
6. Petiole 1–3.5(–5) cm long, fruit sharply 4-angulate, winged lengthwise along the valves ... ... **S. alata**
6. Petiole ca. 3.5–12 cm long; fruit plano-compressed, not winged ... ... **S. reticulata**

Senna alata (L.) Roxb., Fl. Indica 2: 349 (1824). Fig. 1H.
Cassia alata L., Sp. Pl.: 378 (1753).

Woody shrublet to 1 m; stipules ± triangular or lanceolate-ovate, auriculate-amplexicaul on side furthest away from petiole; leaves pinnate, without petiole gland; leaflets bluish green, in 7–14 pairs, broadly oblong to obovate, obtuse to emarginate at apex, the upper face glabrescent, the lower face sparsely pubescent with hairs mainly on the venation; inflorescence a many-flowered raceme; sepals petaloid, deep yellow or orangish-yellow; petals yellow; fertile stamens 2, these with anthers much larger than those of the other stamens; fruit drying black, sharply tetragonal, winged lengthwise down the middle of each valve, the crenate papery wings ca. 4–9 mm wide, many-seeded, the seeds in membranous seed cavities.

Campo.

A pantropical weed that may well have been introduced onto the island after 1978.

Near base-camp, Lewis *1468* (fl. & fr.)!

Senna bacillaris (L.f.) Irwin & Barneby in Mem. New York Bot. Gard. 35(1): 111 (1982).
Cassia bacillaris L.f., Suppl. Pl. 231 (1781).

Only the typical variety has been found on the island.

var. **bacillaris**

Shrub to 2 m; stipules early deciduous, ± linear; leaves pinnate, a pyramidal or dome-shaped, reddish purple gland between the lowest pair of leaflets; leaflets in 2 pairs, their blades subcoriaceous, obliquely elliptic to ovate, discolourous, very finely pubescent on both faces with appressed hairs (these obscure even when viewed with a × 10 lens), tertiary venation prominent, particularly above; inflorescences axillary racemes with the expanded flowers subcorymbose, these compounded into an expansive, terminal pseudopanicle; flowers large and showy, up to 7 cm across; sepals green; petals lemon yellow, finely appressed-pubescent on both faces; functional stamens 7, the three lower ones with a more pronounced bent 'bottle-neck' apex; fruit pendulous, stipitate, cylindrical or obtusely 4-angled (when fully ripe), dehiscent, the valves gaping to expose seeds embedded in a foetid pulp.

Riverside vegetation.

Furo de Maracá, *Milliken* 746 (fl.)!

Senna macrophylla (Kunth) Irwin & Barneby in Mem. New York Bot. Gard. 35(1): 137 (1982).

Cassia macrophylla Kunth, Mimoses pl.: 38 (1822) and: 126 (1823).

One variety occurs on the island.

var. **gigantifolia** (Britton & Killip) Irwin & Barneby in Mem. New York. Bot. Gard. 35(1): 139 (1982).

Chamaecrista gigantifolia Britton & Killip, Ann. N.Y. Acad. Sci. 25(3): 171 (1936).

Shrub 1.5–2 m; stems, petiole, leaf and inflorescence rachises and lower face of leaflets with a fine subappressed to spreading whitish pubescence; stipules linear-setiform, pubescent, persistent (but very variable for the taxon, fide Irwin & Barneby); leaves pinnate; leaflets in 2 pairs with a pyramidal to dome-shaped, purplish black, sessile gland between the basal pair, their blades obliquely broadly ovate to elliptic, up to 30 cm long, acuminate, the upper face glabrous; inflorescence an axillary raceme, much shorter than the leaf that subtends it, the expanded flowers subcorymbose; sepals green drying a dull purplish colour, glabrous, with 5–7 prominent veins arising from the base; petals dull pale yellow, the darker veins very evident on drying, these sparsely pubescent on the outer face; functional stamens 7; fruit pendulous, stipitate, somewhat dorsoventrally compressed and sulcate along the sutures, the valves becoming thin-textured, not known to be dehiscent.

Terra firme forest.

Forest trails near base-camp, *Hopkins et al.* 629 (fl.)!; Proctor Plot 1 [near base-camp], *Milliken* 75 (fl.)!

Senna multijuga (L.C. Rich.) Irwin & Barneby in Mem. New York Bot. Gard. 35(2): 492 (1982).

Cassia multijuga L.C. Rich., Actes Soc. Hist. Nat. Paris 1: 108 (1782).

Only the typical variety occurs on the island.

var. **multijuga** Fig. 1E.

Much branched bushy shrub to small tree, 4–8 m (much taller elsewhere), sometimes multiple-trunked, max. diam. of 5m tree 10cm; bark smooth, blotched shades of grey and white; stipules filiform to narrowly lanceolate, falcately incurved, 4–12 mm long; leaves pinnate, multi-foliolate; a conical gland between the basal leaflet pair; leaflets oblong to oblong-elliptic, mucronate, discolourous, the upper face a glaucous bluish green, the lower face paler, both glabrous or almost so; inflorescences axillary or terminal racemes usually aggregated into showy, leafy panicles; sepals subpetaloid at maturity, yellow, greenish or brownish; petals yellow, heteromorphic; functional stamens 6–7, heteromorphic; fruit a compressed, dehiscent, glabrous, stipitate pod, up to ca. 15 × 2 cm, the shiny reddish or purplish brown valves with a fine transverse venation.

Remnant forest pockets along trackside; secondary vegetation.

Estrada from base-camp to river, *Lewis* 1424 (fl. & fr.)!; near base-camp, *Lewis* 1662 (fl.)!; estrada from base-camp to river, *Ratter et al.* 5493 (fl. & fr.)!

Senna obtusifolia (L.) Irwin & Barneby in Mem. New York Bot. Gard. 35(1): 252 (1982).
Cassia obtusifolia L., Sp. Pl.: 377 (1753).

Weedy herb to 50 cm; stipules linear-lanceolate, incurved to erect, deciduous; leaves pinnate with (2–)3 pairs of leaflets, a peg-shaped gland between the basal leaflet pair; leaflets mid-green, membranaceous, malodorous, broadly obovate, both faces glabrous (on Maracá material), the margins ciliolate; inflorescences 1–2-flowered in the axils of upper leaves, each shorter than the leaf that subtends it; sepals pale green; petals yellow; functional stamens 7; fruit a straight or falcate, narrowly linear, compressed tetragonal, dehiscent pod, two prominent parallel ribs running down the length of the valves.

Disturbed ground.

A pantropical weed, certainly introduced onto the island, probably inadvertently so by man.

Near base-camp, *Lewis* 1695 (fl. & fr.)!

Senna reticulata (Willd.) Irwin & Barneby in Mem. New York Bot. Gard. 35(2): 458 (1982). Fig. 1F.
Cassia reticulata Willd., Enum. Pl. Hort. Reg. Berol. 1: 443 (1809).

Woody shrub to 3 m (elsewhere sometimes a small tree); stipules obliquely triangular-lanceolate, the base auriculately amplexicaul on the side furthest from the petiole; leaves pinnate without petiolar gland; leaflets in ca. 7–13 pairs, slightly discolourous, bluish green on upper face, paler below, oblong to obovate, very sparsely pubescent above, moderately pubescent below; inflorescences at first axillary racemes, later forming exserted candelabriform panicles, many-flowered, in early anthesis capped by a blunt cone of bright yellow or orange-yellow bracts; sepals subpetaloid; petals glabrous, yellow drying whitish; functional stamens 2, the anthers of the 4 median stamens sterile or almost so; fruit a plano-compressed, stiffly papery, stipitate, dehiscenet pod, drying black, ca. 10–15 × 1.3–1.75 cm, many-seeded; seeds hanging from pod on thin funicles after dehiscence, dark olive green outside pleurogram, brownish within.

Sandy banks of igarapé.

On the specimen cited below ants were patrolling the young foliage.

North side of island at Três Igarapés, ca. 1 km up igarapé from sub-camp, *Lewis* 1559 (fr.)!

Senna silvestris (Vell.) Irwin & Barneby in Mem. New York Bot. Gard. 35(1): 87 (1982).
Cassia silvestris Vell., Fl. Flum. 169 (1825).

Only the typical variety has been found on the island.

var. **silvestris**

Small multiple-trunked tree, 4–5 m, thickest trunk ca. 7 cm diam.; young stems longitudinally ribbed and moderately golden-yellow appressed pubescent; stipules linear-setiform, incurved, early caducous; leaves pinnate with ca. (5–)7–11 pairs of leaflets, without petiolar gland; leaflets ovate-lanceolate to ovate-elliptic, acute to acuminate, the upper face glabrous with a fine, reticulate tertiary venation, ± shiny, the lower face pubescent; inflorescences axillary racemes, the expanded flowers subcorymbose, the racemes sometimes aggregated into showy, terminal panicles; sepals subpetaloid, yellowish; petals bright yellow; functional stamens 7, the anthers of 3 longer than those of the median 4; pod as for *S. reticulata* but larger, often over 20 × 2 cm.

Forest margins.

Igarapé Aracacá, *Milliken & Bowles* 259 (fl.)!; Trail 1 (near base-camp), *Ratter et al.* 6322 (fr.)!

4. **CHAMAECRISTA** [Breyne] Moench

Irwin & Barneby in Memoirs N.Y. Bot. Gard. 35(2): 636–892 (1982) (a synoptical revision of the New World species), and in other works cited in their bibliography.

ca. 240 species native to the Americas.

1. Leaves pinnate; gland(s) on petiole stipitate, pinhead-shaped; inflorescence much shorter than leaf that subtends it; stipules not particularly conspicuous, their bases truncate ... ... **C. nictitans** var. **disadena**
1. Leaves bifoliolate; gland on petiole sessile or lacking; peduncle slightly shorter to slightly longer than leaf that subtends it; stipules conspicuous, papery, held erect, their bases ± auriculate:
   2. Gland on petiole sessile; mature leaflets under 25 × 7 mm, their secondary venation ± parallel; plant of the savanna ... ... **C. diphylla**
   2. Gland on petiole lacking; mature leaflets well over 30 × 15 mm, their secondary venation reticulate; plant growing on rocks in centre of river ... ... **C. rotundifolia**

Chamaecrista diphylla (L.) Greene in Pittonia 4: 28 (1899). Fig. 1K.
Cassia diphylla L., Sp. Pl.: 376 (1753); *Cassia cultrifolia* Kunth (1823).

Single-stemmed or sparsely branched, erect, monocarpic or short-lived perennial herb to ca. 60 cm; stipules foliaceous, striate, persistent, held erect, longer than the internodes and thus overlapping, their bases cordate, their apices acute; petiole with a sessile ± round nectary; leaflets in one pair, slightly longer than stipules, coriaceous, prominently veined; flowers single in leaf axils; pedicels at anthesis longer than the leaves that subtend them, pubescent with long, spreading white hairs; petals yellow; fruit a compressed, dehiscent pod, ca. 30 × 5 mm, the valves pubescent, ± 10–15-seeded.

Open savanna on white sand.

Estrada from base-camp to Santa Rosa, *Lewis* 1665 (fl. & fr.)!; Santa Rosa, *Milliken* 370 (fl.)!

Chamaecrista nictitans (L.) Moench, Meth. pl. hort. bot. Marburg, 272 (1794).
Cassia nictitans L., Sp. Pl.: 380 (1753).

Only one variety of this complex species has been recorded from the island.

subsp. **disadena** (Steudel) Irwin & Barneby in Mem. New York Bot. Gard. 35(2): 825 (1982).

Cassia disadena Steudel, Flora (Regensb.) 26: 760 (1843); *Cassia stenocarpa* Vog. (1837); *Chamaecrista stenocarpa* (Vog.) Standley (1916).

var. **disadena** Fig. 1L.

Shrub to 2.5 m; stems pubescent with spreading hairs; stipules striate, acuminate, persistent; leaves pinnate; petiole supporting 1–2 pin-head-shaped, stipitate nectaries; leaflets glaucous bluish green, the largest ones near the centre of the leaf rachis, unlike the stipules in shape and size, the midvein ± central; inflorescence supra-axillary, few-flowered, much shorter than the leaves; petals yellow; fruit a compressed, dehiscent pod, ca. 45 × 3.5 mm, the valves pubescent with short hairs (hardly visible to the naked eye), up to 25-seeded.

Savanna.

Near estrada from base-camp to Rio Urariquera, *Lewis* 1658 (fl. & fr.)!

Chamaecrista rotundifolia (Pers.) Greene in Pittonia 4: 31 (1899). Fig. 1M.

Cassia rotundifolia Pers., Syn. Pl. 1: 456 (1805).

Woody-based herb or shrublet to 80 cm; stems tinged red, sparsely pubescent; stipules as for *C. diphylla* but papery and not or hardly overlapping, especially on older stems; petiole lacking a nectary; leaflets in one pair, yellowish green, obovate to broadly oblanceolate, several times larger than stipules, strongly asymmetric about the midvein, rounded to retuse at apex, glabrous but for the ciliolate margins; flowers single in leaf axils, the pedicel ± equalling the leaflets in length, glabrous; petals yellow; fruit a dark reddish brown, compressed, dehiscent pod, ca. 50 × 5 mm, the valves very sparsely pubescent, ± 8–12-seeded.

River smoothed rocks amongst white sand in centre of river.

Furo de Maracá heading west from base-camp, *Lewis* 1670 (fr.)!; same locality, *Milliken et al.* 424 (fl. & fr.)!

5. **BAUHINIA** L.

About 250–300 spp., pantropical.

1. Shrub or spindly treelet, 2–6 m tall; flower buds cylindrical, several times longer than wide; fruit over 12 cm long, ca. 1 cm wide, more than 10-seeded; characteristic of forest-savanna boundary ... ... **B. ungulata**
1. Woody climbers; flower buds globose to ovoid, slightly longer than wide; fruit up to 11 cm long, 3–6-seeded; plants of riverine or terra firme forest:
   2. Sepals expanded, ca. 2 mm wide, the 5 together forming a 'crown' ± equalling or wider than the calyx tube in bud; mature fruits rusty brown tomentose ... ... **B. outimouta**

Fig. 1., all × ⅔. **A**: *Elizabetha coccinea* var. *oxyphylla*, **i**, before dehiscence (from Lewis 1673), **ii**, single valve after dehiscence (from Hopkins et al. 741); **B**: *Bauhinia glabra* (from Lewis 1430); **C**: *Macrolobium acaciifolium* (from Henderson & de Lima 611); **D**: *Peltogyne paniculata* subsp. *pubescens* (from Lewis 1606); **E**: *Senna multijuga* var. *multijuga* (from Lewis 1424); **F**: *Senna reticulata* (from Lewis 1559); **G**: *Bauhinia ungulata* (from Lewis 1462); **H**: *Senna alata* (from Lewis 1468); **J** *Cassia moschata*, **i**, under view, **ii**, side view (from Lewis 1423); **K**: *Chamaecrista diphylla* (from Lewis 1665); **L** *Chamaecrista nictitans* var. *disadena* (from Lewis 1658); **M**: *Chamaecrista rotundifolia* (from Lewis 1670); **N**: *Hymenaea parvifolia* (from Lewis 1421).

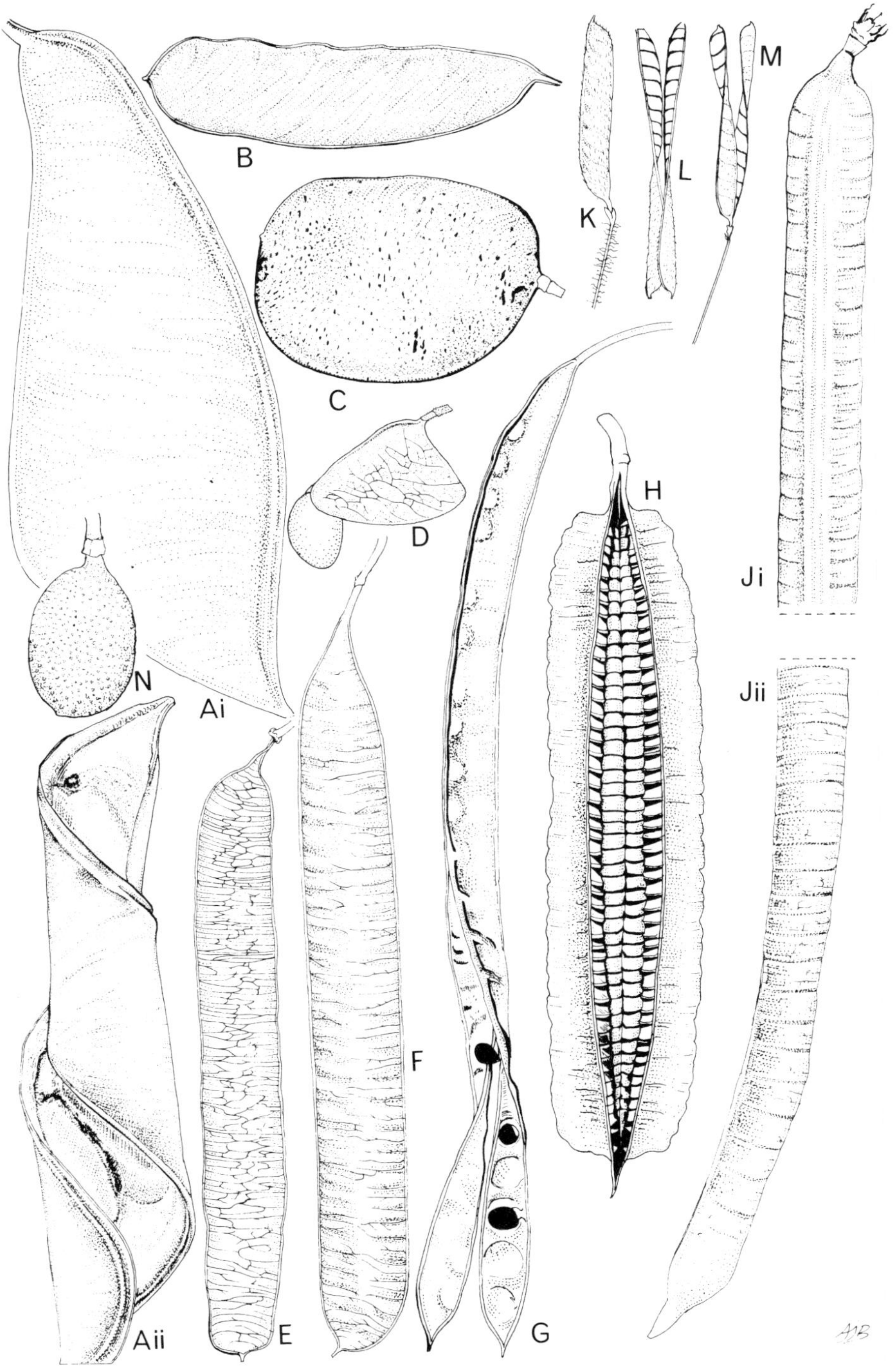
Ai
Aii
B
C
D
E
F
G
H
Ji
Jii
K
L
M
N

2. Sepals minute or linear-lanceolate, acuminate, under 1 mm wide; mature fruits glabrescent:
   3. Sepals minute, under 1 mm long; mature leaves bullate, their bases subcordate ... ... **B. guianensis**
   3. Sepals linear-lanceolate, ca. 2–3 mm long; mature leaves not bullate, their bases deeply cordate to auriculate ... ... **B. glabra**

Bauhinia glabra Jacq., Enum. Pl. Carib.: 20 (1760). Fig. 1B, Plate 2C.

Woody climber reaching ca. 8 m, older stems flattened, younger stems terete, clothed in a rusty-red tomentum; bark of previous years growth flaking off in thin papery strips, younger branchlets with many simple, velvety tendrils; leaves unifoliolate, deeply bilobed to ca. $\frac{1}{3}$ to $\frac{1}{2}$ blade length, deeply cordate at the base especially on older leaves, discolourous, shiny dark green and glabrous on upper face, paler on lower face, the latter with a rusty-red tomentum, especially on the veins of young leaves, glabrescent at maturity, the hairs then sometimes whitish; inflorescence a terminal raceme, the long acuminate calyx lobes and bracts forming a prominent tuft at the apex of the young inflorescence; petals white; fruit a dehiscent, stipitate pod, drying black and shiny, ± 4-seeded; seeds dark olive brown, round, compressed.

Terra firme forest and riverine forest (but not in areas prone to flooding).

Rio Urariquera, between base-camp and Maracá house, *Lewis* 1430 (fl. & fr.)!; north side of island at Três Igarapés, ca. 1 km up Igarapé from sub-camp 1, *Lewis* 1561 (fr.)!; along edge of Furo de Maracá, *Henderson et al.* 566 (fl.)! (NY, photo K); Furo de Maracá near eastern end of island, *Sette Silva & Lima* in *Ratter* 5819 (fr.)!

Bauhinia guianensis Aubl., Pl. Guian. Fr. 1: 377, t.145 (1775).

Robust woody liana reaching forest canopy, older stems near forest floor flattened and strap-like, up to 15 cm wide, undulating, occasionally with a short woody tendril on the raised portions, longitudinally ridged; leaves unifoliolate, sometimes very large, up to 25 × 25 cm, coriaceous, bullate, discolourous, the upper face deep bottle green, shiny, glabrous, lower face ferruginous tomentose, with age sometimes glabrescent, deeply bilobed, often to the petiole on leaves produced on older parts of plant, the two lobes acute to acuminate apically, usually producing drip-tips; inflorescence a terminal raceme to ca. 15 cm long; petals white; fruit a dehiscent flattened pod to ca. 9 cm long, reddish brown tomentose, stipitate, 3–5-seeded.

Terra firme forest; open forest on steep slopes in W. of island.

100 m along estrada from base-camp to Santa Rosa, *Lewis* 1603 (sterile)!; western tip of island near Pururamé falls, *Lewis & Milliken* 1686 (fruit and seedlings)!; low ridge south of the Cachoeira de Fumaça, *Milliken et al.* 495 (sterile)!

Bauhinia outimouta Aubl., Pl. Guian. Fr. 1: 375 (1775).
Bauhinia rubiginosa Bong. (1838).

Woody liana reaching forest canopy although often flowering on lower stems, young stems clothed with a rusty brown tomentum; leaves unifoliolate but deeply bilobed, usually to the petiole, coriaceous, strongly discolourous, the upper face bluish green, glabrous, lower face with a dense rusty-brown tomentum; inflorescence terminal or axillary, racemose or occasionally paniculate; calyx lobes expanded, broadest towards apex, the 5 together forming a 'crown' equalling or wider than the calyx tube in bud; petals creamy white on their inner face, golden yellow on outer face due to dense silky pubescence; stamen filaments white; anthers yellow; fruit a dehiscent, oblong, flattened pod to 11 cm long, ferruginous-tomentose, stipitate, 5–6-seeded.

Terra firme forest; low forest with abundant palms.

Edge of base-camp clearing, *Hopkins et al.* 672 (fl)!; same general locality, *Milliken* 180 (fl.)!; same locality, *Lima* 802 (fl.)!

Bauhinia ungulata L., Sp. Pl.: 374 (1753). Fig. 1G, Plate 2B.
Bauhinia macrostachya Benth. (1840), non Wall. ex Bak. (1878); *Bauhinia benthamiana* Taub. (1891).
"Mororó", "Unha de vaca".

Shrub to spindly treelet, 2–6 m (DBH of 6 m treelet 10 cm); leaves unifoliolate, bilobed, the blade divided for ⅓ to ⅔ its length, coriaceous, dark bottle green, the upper face glabrous, the lower face sparsely pubescent with very short hairs, with prominent reticulate tertiary venation; inflorescence a terminal raceme; buds sausage-shaped; calyx lobes reflexing to expose very narrow whitish petals; stamen filaments turning from white to red with age; fruit an explosively dehiscent, woody, elongate pod, 10–25 cm long, the valves twisting after dehiscence to expose up to ca. 25 seeds; seeds dark brown.

Campo — savanna margin; low campina forest; *Curatella americana* savanna; savanna — forest transition.

Of the species of *Bauhinia* recorded for Maracá this is the only one that is not a woody climber. It is common in many parts of Brazil and has been collected on the road from the island to Boa Vista. On Maracá it is very evident on the estrada from base-camp to the Furo de Maracá and is one of the legume species characterizing the forest — savanna boundary near base-camp.

Estrada from base-camp to Rio Urariquera, *Lewis* 1426 (fl. & fr.)!; campo — cerrado margin by base-camp, *Lewis* 1462 (fl. & fr.)!; margin of wet campo just below base-camp, *Ratter et al.* 5357 (fl. & fr.)!; Santa Rosa campo, *Ratter et al.* 5691 (fl.)!; Santa Rosa, *Ratter et al.* 5691 (fl.)!; Santa Rosa, *Ratter et al.* 5718V (sterile)!

6. **PELTOGYNE** Vog.

Freitas da Silva in Acta Amazonica 6(1), suppl.: 1–61 (1976) (revision).
23 spp. Mexico and W. Indies to SE. Brazil, mostly in Amazonia.

Buds globose, moderately appressed-pubescent; ovary glabrous; sepals early caducous; fruit sublunate, narrowing towards the short, ca. 1 mm long, stipe ... ... **P. gracilipes**
Buds turbinate, densely appressed white pubescent; ovary woolly; sepals persisting as ovary develops, reflexed, a dense woolly patch on the centre of the inner face; fruit subtriangular, broadening towards the 2–3 mm long stipe ... ... **P. paniculata** subsp. **pubescens**

Peltogyne gracilipes Ducke in Trop. Woods 54: 6 (1938). Plate 3D.

Tree, 12–30 m, DBH 30–67 cm, sometimes with buttresses (e.g. *Milliken et al.* 100V), crown with open branching; bark greyish white, smoothish, mottled due to falling chunks; slash pink, odourless, no evidence of exudate; leaves bifoliolate; leaflets coriaceous, falcate, strongly asymmetric about the midvein, acute to acuminate at apex (the drip-tips particularly long on seedling leaflets), discolourous, dark green above, paler below, glabrous, both faces with a finely reticulate, prominent tertiary venation; inflorescence an axillary or terminal panicle; buds globose; rachises, peduncles, pedicels and sepals pubescent; petals white; ovary glabrous, shortly stipitate; stigma capitate; fruit a subligneous, glabrous, dehiscent pod, ca. 3–3.5 × 1.5–1.8 cm, the upper margin straight, the lower one curved, the body of the fruit sublunate, narrowing towards the short stipe, a prominent, reticulate venation especially evident towards the apex.

Riverine forest, especially on white sand; seasonally flooded várzea forest.

Furo de Maracá, *Lewis* 1653 (fr.)!; near Casa de Maracá, *Milliken et al.* 100V (sterile)!; Ilha de Nova Olinda, *Milliken & Bowles* 399 (fl.)!; *Ratter et al.* 5893V!, a sterile collection from close to the Igarapé camp, Fumaça trail and *Milliken et al.* 159V from near the Casa de Maracá, may also belong here.

Peltogyne paniculata Benth. in Hook. Journ. Bot. 2: 96 (1840).
One subspecies is known from the island.

subsp. **pubescens** (Benth.) M.F. da Silva in Acta Amazonica 6(1), suppl.: 38 (1976). Fig. 1D, Plate 3C.
Peltogyne pubescens Benth. in Hook Journ. Bot. 2: 96 (1840).
"Pau roxo", "Roxinha", "Violeta".

Tree, 3–25 m, DBH 10–40 cm, sometimes buttressed and sometimes ± multiple-trunked; bark dark greyish black or yellowish brown with darker lenticels in horizontal rows, smoothish; slash yellow to reddish, no exudate, wood drying a characteristic mauve-violet; leaves bifoliolate; leaflets coriaceous, oblong to oblong-elliptic, obtuse to rounded or acute at apex, yellowish green, asymmetric about the midvein especially at the base, glabrous; inflorescence an axillary or terminal panicle, most parts densely woolly with white hairs; buds turbinate; hypanthium and reflexed sepals pinkish; petals 4, pale pink, soon falling; stamen filaments white; style pink; ovary moderately to densely pubescent with white hairs; fruit a subtriangular, straw-coloured, dehiscent pod, the subligneous endocarp separating to leave a single reddish brown seed attached to the apex of one coriaceous valve, thus forming a pseudosamara, the valves glabrous or almost so, shiny, with a prominent reticulate venation.

Riverine forest; terra firme forest; low campina forest; savanna; transition between forest and campo.

The vernacular name "pau roxo" alludes to the characteristic mauvish violet wood. The single seed firmly attached to the apex of one fruit valve, thus producing a pseudosamara, is dispersed by wind with the whole structure floating downwards like a slow-moving helicopter blade.

Between base-camp and Maracá House, *Lewis* 1437 (fr.)!; estrada from Santa Rosa to base-camp, *Lewis* 1449 (sterile)!; Furo de Maracá, *Lewis* 1674 (fl.)! forest north of Casa de Maracá, *Milliken* 117V (sterile)!; Ilha de Nova Olinda, *Milliken & Bowles* 398 (fl.)!; Santa Rosa, *Ratter et al.* 5720V (sterile)!; Proctor Gap plot [near base-camp], *Ratter et al.* 5751V (sterile)!; Santa Rosa Campo, *Ratter et al.* 5557 (fr.)!; Trilha de Preguiça, 2–3 km N. of Casa de Maracá, transect tree B15/2, *Ratter et al.* 6221V (sterile)!; ca. 4 km NE. of Santa Rosa Station buildings, *Ratter et al.* 6330 (fr.)!; estrada from base-camp to Santa Rosa, *Ratter et al.* 6335 (fr.)!

Ratter et al. 5625, a sterile collection of juvenile foliage remains unidentified to species.

7. **HYMENAEA** L.

Lee & Langenheim in Univ. Calif. Publ. Bot. 69: 1–109 (1975) (revision).

15 spp., Mexico, Central and tropical South America, with one species on E. coast of Africa, Madagascar and Mascarene Is.

Inflorescences short-paniculate, 8–15 cm long, corymbose when mature, flower buds ca. 12–20 mm long; petals glabrous; ovary glabrous, subsessile; fruit over 6 × 3 cm, ca. 4–6-seeded ... ... **H. courbaril**
Inflorescences long-paniculate, usually over 18 cm long; flower buds smaller; petals pilose on outer face with rigid white hairs; ovary pubescent over whole surface, stipitate; fruit up to 5 × 2.8 cm, 1-seeded ... ... **H. parvifolia**

Hymenaea courbaril L., Sp. Pl.: 1192 (1753).

Available material from Maracá is not adequate for placement at varietal rank although the collection cited below most probably represents var. **subsessilis** Ducke.

Trees 8–20 m, DBH 25–35 cm; bark greyish black; leaves bifoliolate; leaflets coriaceous, glabrous, shiny, the blade asymmetric about the midvein, the apex acute to short acuminate; inflorescence a compact, terminal panicle becoming densely corymbose at anthesis; buds pale green; calyx lobes tomentose on both faces, the inner face with a central, vertical pilose area; petals creamish; ovary glabrous, subsessile, sitting in a short woody hypanthium; fruit (not known from Maracá, but see note below) cylindrical, subcompressed, indehiscent, 4–6-seeded.

Terra firme forest.

Some collections of *Hymenaea* from Maracá are made up of loose fruits and flowers collected from the forest floor together with foliage taken from nearby trees. Given that the two species recorded from the island occur in the same patch of forest between base-camp and Santa Rosa it is possible that some collections may be mixed gatherings. This is possibly the case with *Milliken & Miller* 822 which has fruits similar to those of *H. courbaril* var. *subsessilis*, foliage that seems to match collections of both *H. courbaril* sens. lat. and *H. parvifolia* and flowers which match neither, the petals not being pilose (on the inner face) unlike the distinctly stipitate ovary which is. Based on flower morphology differences, a restricted area of forest seems to support three separate taxa, but equally probable is that *Milliken & Miller* 822 is of hybrid origin, the putative parents being *H. courbaril* and *H. parvifolia.* The latter suggestion supports Miller's field observation (pers. comm., Milliken 1988).

Estrada from base-camp to Santa Rosa, *Lewis* 1700 (fl.)!;

Hymenaea parvifolia Huber in Bol. Mus. Paraense Hist. Nat. 5: 385–6 (1909). Fig. 1N, Plate 3A & B.
"Jatobá"

Trees, 20–40 m, sometimes shrub-like, DBH 30–40 cm; bark grey to pale brown, smooth; slash deep pink; wood hard; leaves bifoliolate; leaflets coriaceous, glabrous, shiny, the blade asymmetric about the midvein, the apex acute or rounded, sometimes acuminate (drip-tipped) on seedlings; inflorescence paniculate; buds green; calyx lobes tomentose on outer face, white pilose on inner face; petals creamy white, pilose on outer face with rigid white hairs; stamen filaments creamy white; ovary stipitate, pubescent over whole surface; fruit ovoid to obovoid, smooth, relatively small for the genus, ca. 3.5–5 × 2.2–2.8 × 2–2.5 cm, indehiscent, brown with paler warty speckling, mesocarp pale green, usually 1-seeded.

Terra firme forest.

Estrada from base-camp to Santa Rosa, *Lewis* 1421 (fl. & fr.)!; regeneration plots [near base-camp], *Milliken & Miller* 771 (fl. & fr.)!; ca ½ km from Santa Rosa station, *Ratter et al.* 5464V (fr.)!; Proctor gap plot [near base-camp], *Ratter et al.* 5772 (fr.)!

8. **ELIZABETHA** Schomb. ex Benth.

Cowan in Konikl. Nederl. Akad. Wetensch. Amster., series C, 79(4): 323–346 (1976) (Revision of the genus; includes keys).
11 spp., tropical South America.

Elizabetha coccinea Schomb. ex Benth. in Hook. Journ. Bot. 2: 92 (1840).
One of the two described varieties occurs on the island.

var. **oxyphylla** (Harms) Cowan in Konikl. Nederl. Akad. Wetensch. Amster., series C, 79(4): 340 (1976). Fig. 1A, Plate 2 E & F.
Elizabetha oxyphylla Harms in Notizbl. 6: 306 (1915).

Shrub to moderate-sized tree, 3–10 m, DBH of 10 m tree 10 cm (can attain 20 m with a 20 cm diam. trunk, fide Cowan); stipules early caducous; leaves pinnate with (3–)6–8(–12) pairs of leaflets, the rhachis channelled along the upper edge; leaflets yellowish green, sessile, ± coriaceous, elliptic to oblong-elliptic or obovate, the bases inequilateral, distinctly lobed on one side, otherwise the blade divided ± 50/50 by the midvein, the blades essentially glabrous but for a few hairs on the midvein and occasionally the margins, tertiary venation evident on both faces; inflorescence a terminal raceme, 10–20(–30) cm long; bracts large and showy, up to 5 × 3 cm; bracteoles smaller, joined for ½ to ¾ their length; hypanthium greenish yellow, subligneous, to 3 cm long, glabrous; calyces creamish; petals 5, scarlet; stamen filaments blood-red; ovary deep crimson; fruit a stipitate, woody, explosively dehiscent pod, yellowish green tinged red, maturing golden orangish brown due to a dense felty tomentum, oblong-oblanceolate, ca. 12–20 × 4–6 cm, the valves obscurely transverse-veined, the upper margin 'winged', the wing 4–5 mm wide, ca. 4–6-seeded; seeds laterally compressed, the thin testa separating from the fleshy cotyledons on drying, orangish brown with an incompletely reticulate venation of prominent chocolate-brown veins which fail to join up at the testa centre (giving the dried seed the appearance of a miniature lung).

Riverine forest.

A common species especially along the Furo de Maracá at the eastern end of the island. The leaflets are often held ± erect and together with the very showy inflorescence make the species easy to recognise in dense riverine forest. *Henderson et al.* 604 report that the flowers were visited by numerous trigonid bees. *Milliken & Bowles* 290, a specimen of *E. princeps* Schomb. ex Benth., was collected up-river from Maracá but is to be expected at the western end of the island.

Furo de Maracá, riverine islet in midstream, *Lewis* 1673 (fl. & fr.)!; along edge of Furo de Maracá, *Henderson et al.* 604 (fl.)!; Furo Pananá de Firmino of Rio Urariquera on south side of island, *Hopkins et al.* 741 (fl. & fr.)!; Ilha Nova Olinda, *Milliken & Bowles* 389 (fr.)!

9. **MACROLOBIUM** Schreb.

Cowan in Mem. N.Y. Bot. Gard. 8(4): 257–342 (1953) (revision).
About 60 spp. in tropical America.

Macrolobium acaciifolium (Benth.) Benth. in Mart. Fl. Bras. 15(2): 224 (1870). Fig. 1C, Plate 2D.
Outea acaciaefolia Benth. in Journ. Bot. Hook. 2: 94 (1840).

Tree (3–)8–15 m, DBH 15–30 cm, sometimes multiple-trunked with an elegant rounded crown; bark whitish grey to greyish brown, rough and fissured; leaves pinnate, the rachis slightly winged between each pair of opposite leaflets; leaflets ± oblong, obtuse to emarginate at apex, inequilateral at base, coriaceous, glabrous but for a few hairs on the mid vein below, discolourous, the upper face dark green, the lower face paler; inflorescence an axillary raceme much shorter than the leaf that subtends it, usually held ± erect, most parts with a dense whitish yellow pubescence; bracts and calyces pale greenish cream; the solitary petal white; stamen filaments white for lower ⅓, crimson for upper ⅔; fruit a tardily dehiscent woody pod (often floating some distance from source tree before rotting and opening on sandy river banks), yellowish green, maturing orangish light brown, up to ca. 7 × 5 cm.

Riverine forest.

A frequently encountered species along the river margins particularly in the east of the island. The woody fruits are commonly used in the Amazon by the local Indians to make necklaces and decorative wall hangings.

Furo de Maracá, cachoeira up-river from Casa de Maracá, *Lewis* 1535 (fr.)!; Furo de Maracá, *Lewis* 1676 (fl. and fr.)!; same genera locality, *Lewis* 1677 (fr.)!; same locality, *Henderson et al.* 611 (fr.)!; near cachoeira de Prumama [= Pururamé], Furo de Santa Rosa, *Milliken et al.* 715 (fr.)!

10. **PENTACLETHRA** Benth.

2 spp., Central America, northern South America, tropical Africa.

Pentaclethra macroloba (Willd.) O. Kuntze, Rev. Gen. Pl. 1: 201 (1891). Fig. 2A.
Acacia macroloba Willd., Sp. Pl. 4: 1060 (1806); *Pentaclethra filamentosa* Benth. (1840).

Tree 7(–25) m, DBH of 7 m tree 25 cm, trunk buttressed, crown dense, broad; leaves bipinnate with 10–20 pairs of pinnae; leaflets in ca. 25–50 pairs, coriaceous, glabrous, asymmetric about the midvein especially at the base, acute to apiculate at apex; inflorescence an axillary or terminal spike, the spikes sometimes aggregated into a panicle, floral scars very evident on the tomentose rachis after flowers have fallen; corolla red externally, yellow internally; fruit an explosively dehiscent woody pod with the valves coiling up after dehiscence, ca. 30–40 × 3.5–6 cm, falcate, longitudinally veined, ca. 6-seeded.

Terra firme forest near river.

The bark is reported to be used as a fish poison (Fruits of the Guianan Flora, 1985).

Furo de Santa Rosa, *Hopkins et al.* 877 (fr.)!

11. **PARKIA** R. Br.

Hopkins, Flora Neotropica monograph 43: 1–124 (1986).
About 40 spp. in tropical South America, Africa and Asia.

Parkia pendula (Willd.) Benth. ex Walp., Rep. Bot. Syst. 5: 577 (1846). Fig. 2K.
Inga pendula Willd., Sp. Pl. ed. 4, 4: 1025 (1806).

Large emergent tree to 35(–50) m with a broad, flattened, layered or umbrella-shaped crown, DBH 80 cm, sometimes buttressed; bark thick, rough and shaggy, orangey brown, flaking in thick woody chunks; outer slash a salmon-pinkish orangey brown colour, inner slash white to creamish yellow, slash releases a very weak smell of shallots (*Allium* spp.); leaves bipinnate; petiole bearing a small circular or elliptical gland on upper face, halfway between base and first pinnae pair; pinnae in 15–27 opposite or alternate pairs, multifoliolate; leaflets opposite, ± linear, ca. 3–6 × 0.5–1.5 mm, subcoriaceous, discolourous, the upper face dark bottle green, nitid, the lower face paler, glabrous but for the sparsely ciliate margins; inflorescence a pendulous compound raceme or panicle of bright red heads, up to 1 m long, the heads ca. 4–5 cm in diam.; bracts at base of capitula green; flower bracts golden-yellow; flowers dimorphic, either hermaphrodite or nectar-secreting; filaments, anthers and style purplish; fruit a strap-shaped, subligneous, glabrous pod, dehiscent along the lower, much thickened, margin which secretes large quantities of sticky amber-coloured gum into which the seeds are released, 15–30 × 2–3 cm, 17–25(–30)-seeded.

Upper slopes of terra firme forest on weathered schist; 150–200 m alt.

Western tip of island near Pururamé falls, *Lewis & Milliken* 1685 (fl.)!; same locality, *Milliken et al.* 642 (fr.)!

12. **ENTADA** Adans.

About 25 spp., widespread, mainly tropical, most numerous in Africa.

Entada polystachya (L.) DC., Mém. Leg. : 434 (1826). Fig. 2B, Plate 3F.
Mimosa polystachya L., Sp. Pl. : 520 (1753); *Entadopsis polystachya* (L.) Britton (1928).
"Rabo de macaco", "Cipó escova".

High-climbing woody liana to 10 m or more and regularly reaching forest canopy; petioles sometimes developed as tendrils; bark grey; leaves bipinnate, pinnae in 2–6 pairs; leaflets in 6–10 pairs, opposite, discolourous, mid bottle-green above, paler below, sparsely to moderately softly pubescent on both faces, glabrescent above; inflorescence terminal, the tiny flowers in spikes aggregated into a massive brush-like structure (looking like an upturned broom head) 50 cm or more in length, at anthesis with a strong foetid odour similar to rotting fish; calyx green; petals pinkish, especially at their bases, often tinged green; stamen filaments white or creamish; anthers yellow; style pink; fruit a true craspedium, ca. 25–30 cm long, the exocarp peeling away in papery flakes, the endocarp breaking up into one-seeded, winged packages and leaving a persistent replum; seeds compressed, orangish brown, ca. 16 × 10 mm, shiny, the pleurogram submarginal, open at apex.

Riverine forest and trackside scrubland.

Entada is a genus having some of the smallest leguminous flowers that produce some of the largest legume fruits. On Maracá in the dry season, river banks are often covered in the one-seeded papery 'packages' that are dispersed by both wind and water. The foul smelling flowers attract numerous insect species.

Between base-camp and Maracá house, *Lewis* 1432 (fl. & fr.)!; 'causeway' between base-camp and river, *Milliken* 63 (fl.)!; riverbank near Casa de Maracá, *Milliken et al.* 168 (fr.)!; as for Milliken 63, *Ratter et al.* 5500 (fl.)!

13. **ANADENANTHERA** Speg.

von Reis Altschul in Contrib. Gray Herb. 193: 1–65 (1964) (revision of the genus including a key to the taxa).

2 spp. in West Indies and northern South America.

Anadenanthera peregrina (L.) Speg., Physis 6: 313 (1923). Fig. 2D.
Only the typical variety occurs on Maracá.

var. **peregrina** Plate 3E.
Mimosa peregrina L., Sp. Pl. : 520 (1753); *Acacia peregrina* (L.) Willd. (1806); *Piptadenia peregrina* (L.) Benth. (1841); *Niopa peregrina* (L.) Britton & Rose (1927).
"Angica", "Angico", "Angico preto".

Tree 5–15 m, DBH 6–20(–35) cm, often flowering and fruiting when leafless; bark greyish, deeply rugose, warty protuberances on trunk and main branches of some specimens, warts occasionally ± spine-tipped; slash yellow; leaves bipinnate, feather-like; an elliptic nectary on the petiole; leaflets tiny, under 5 × 1 mm, yellowish green; flowers in heads, these solitary or grouped into fascicles in leaf axils or ± ramiflorous on present years growth, faintly scented, an involucre of bracts ⅔ up the peduncle; calyx and corolla cream tinged green; stamen filaments white; anthers pale yellow; fruit a brown, scaley dehiscent pod opening along its upper margin, up to 30 cm long, irregularly constricted between the seeds especially along the lower margin; seeds flat, round, shiny, narrowly 'winged'.

Forest margin transition with campo; granitic outcrops with *Cactus* and *Vitex*; on laterite hilltop.

A common species at the east of the island, characteristic of the forest margins around base-camp. The seeds have hallucinogenic properties and have been widely used in Amerindian cultures (see von Reis Altschul, 1972).

SW of base-camp, ca. 750 m to right of estrada down to Rio Urariquera, *Lewis* 1493 (fr.)!; same locality, *Lewis* 1602 (fl.)!; Rio Uraríquera, a 6 km baixando o rio, *Lima et al.* 816 (young fr.)!; near base-camp, *Milliken & Lima* 46 (fl. & fr.)!; below estação, *Ratter et al.* 5568 (fr)!; same locality, *Ratter et al.* 5745 (fr.)!; same locality, *Ratter et al.* 6275 (fl.)!; granite outcrops along Furo de Maracá, *da Silva et al.* 2391 (young fr.)!

14. **MIMOSA** L.

Barneby is about to publish a revision of the South American species.

About 500 to 600 spp., predominantly in Central and South America, but extending thinly to Africa and Asia.

1. Mature leaflets over 10 mm wide; 2–4 leaflets per pinna; leaves consistently with 1 pair of pinnae only:
  2. Woody-based herb; stamens pink; leaflets rounded to obtuse, mucronate at apex; older stems terete, not strongly ribbed although sometimes with scattered, spreading or slightly recurved prickles ... ... **M. debilis**
  2. Scrambling climber; stamens white; leaflets acute at apex; older stems angular, strongly ribbed with numerous reflexed prickles along the ribs ... ... **M. sensitiva**
1. Mature leaflets under 5 mm wide; pinnae multifoliate:

3. Leaves multipinnate; erect spine-like spicules between each pinna pair; fruit ca. 10 mm wide, the narrowly-oblong articles over 10 per fruit **M. pellita**
3. Leaves with 1–2 pairs of pinnae; lacking spicules although sometimes with falcate prickles along lower edge of pinnae rachises; fruit ca. 5 mm wide, the orbicular or squarish articles 5 or fewer per fruit:
   4. Leaves usually with 2 pairs of pinnae; leaflets 1.5–2 mm wide; stems terete, glabrous or glabrescent, armed with spreading ± straight prickles; inflorescence ellipsoidal; articles of fruit glabrous ... ... **M. pudica** var. **tetrandra**
   4. Leaves with 1 pair of pinnae only; leaflets 4–5 mm wide; stems ± angular, densely pubescent, armed with numerous scattered, recurved prickles; inflorescence globose; articles of fruit pubescent with spreading bristles intermixed especially over seed cavities ... ... **M. schrankioides**

Mimosa debilis Humb. & Bonpl. ex Willd., Sp. Pl. 4: 1029 (1806). Fig. 2J.

Woody-based bushy herb or shrublet to 1 m; stems with golden-yellow patent bristles intermixed with a much shorter white crisped pubescence and scattered recurved prickles; stipules lanceolate, acuminate, striate, persistent, ca. 5 mm long, the margins with patent bristles; leaves bipinnate, the pinnae reduced to a single pair, the petiole and pinnae rachises as for the stems although often lacking prickles, each pinna with two pairs of leaflets, one leaflet of the lower pair several times larger than the other and ± equalling the leaflets of the upper pair; leaflets bluish green, strongly asymmetric about the midvein, appressed-pubescent and appressed-setose on both faces or without setae on upper face, sensitive; inflorescences compact pedunculate heads, these often paired in the leaf axils or appearing as a terminal raceme of heads; bracts and bracteoles lanceolate, acuminate, ciliate; flowers creamish green; stamen filaments pale pink; fruit ± a craspedium, the valves separating as one unit to leave a persistent rim (replum), valves with a mixture of short white hairs and golden bristles, the margins with more robust reddish purple bristles.

Secondary vegetation and campo.

Near base-camp, *Lewis* 1472 (fl.)!; estrada from base-camp to river, *Lewis* 1570 (fl. & fr.)!; same locality, *Milliken* 202 (fl. & fr.)!

Mimosa pellita Humb. & Bonpl. ex Willd., Sp. Pl. 4: 1037 (1806). Fig. 2F.

Shrublet, 1.5–2 m, the ribbed stems with a fine white pubescence (hardly visible to the naked eye) with patent yellowish bristles intermixed and scattered, recurved woody prickles; stipules firm, densely appressed-setose on outer face; leaves bipinnate with ca. 10–15 pairs of pinnae, the leaf and pinnae rachises with a fine white pubescence and intermixed bristles, a 3–10 mm long erect woody spicule between each pinnae pair, the pinnae multifoliolate; leaflets bluish green, ± oblong-lanceolate, acute, ca. 4–5 × 1–1.5 mm, ca. 3-costate, the lower face appressed-pubescent, the margins with patent setae, sensitive; inflorescences

Fig. 2., all × ⅔. **A**: *Pentaclethra macroloba*, **i**, before dehiscence (from British Guiana Forest Dept. 2965), **ii**, inner face of valve after dehiscence (from Berti 74); **B**: *Entada polystachya* (from Lewis 1432); **C**: *Acacia articulata* (from Lewis 1527); **D**: *Anadenanthera peregrina* var. *peregrina*, **i**, fruit, **ii**, seed (from Lewis 1493); **E**: *Acacia paniculata* (from Lewis 1501); **F**: *Mimosa pellita* (from Lewis 1571); **G**: *Mimosa sensitiva* (from Lewis 1500); **H**: *Mimosa pudica* var. *tetrandra* (from Lewis 1567); **J**: *Mimosa debilis* (from Lewis 1570); **K**: *Parkia pendula* (from Krukoff 7113).

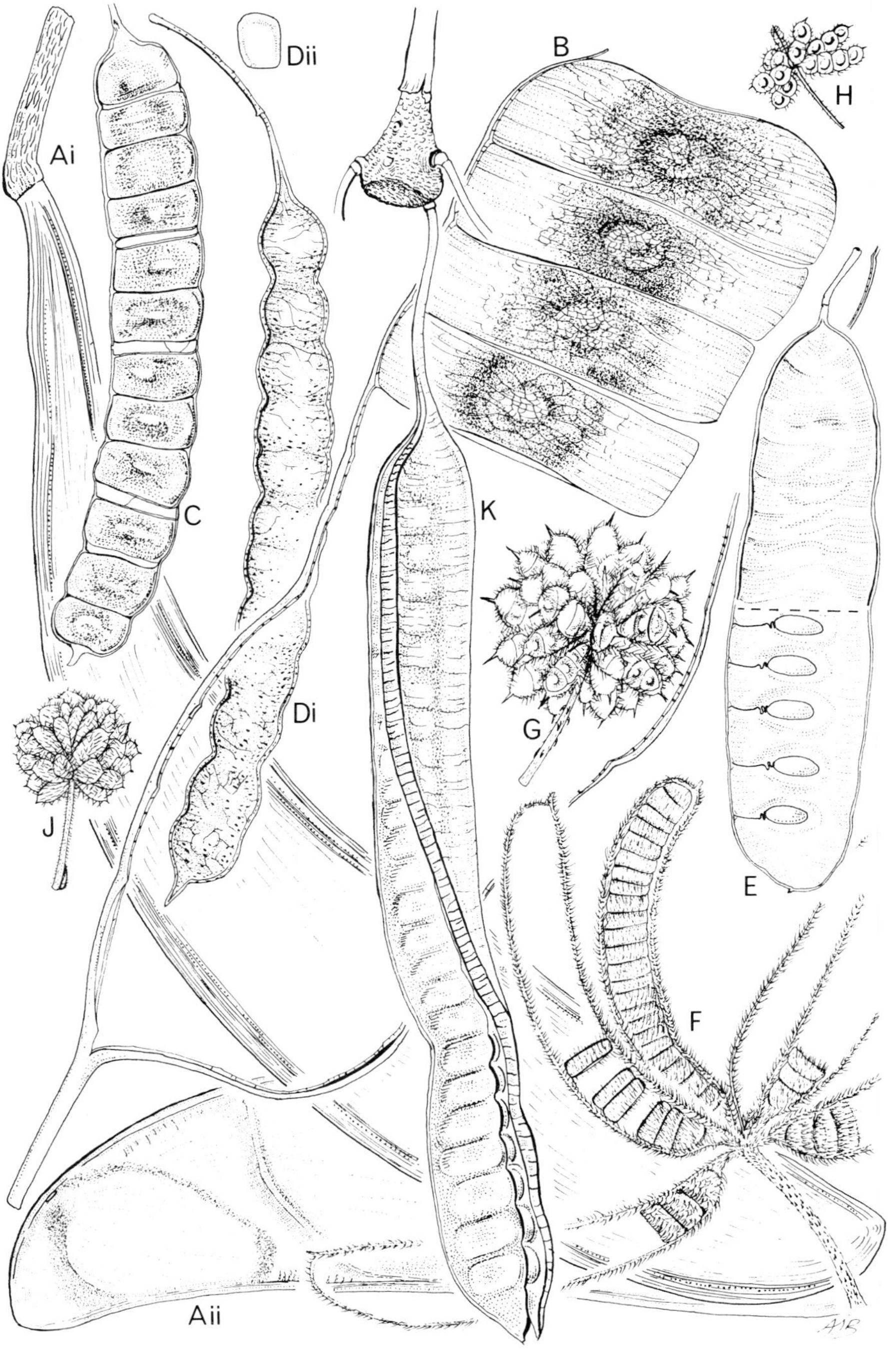
Ai
C
Dii
B
H
K
E
G
Di
J
F
Aii

pedunculate heads, single or clustered in the axils of the upper leaves or aggregated into a terminal leafy raceme of heads; peduncle densely appressed-setose; flowers greenish; stamen filaments pale pinkish; fruit a true craspedium breaking up into 1-seeded articles to leave a persistent replum, the articles and replum with spreading bristles.

Campo; secondary scrub and open savanna.

Recognised as a distinct species within the *Mimosa pigra* L. "complex" by Barneby (pers. comm., 1987). According to Barneby true *M. pigra* is restricted to the Paraguai basin in Paraguay and Argentina.

Near base-camp, *Lewis* 1457 (fl. & juvenile fr.)!; estrada from base-camp to river, *Lewis* 1571 (fl. & fr.)!; near base-camp, *Hopkins et al.* 570 (fl.)!

Mimosa pudica L., Sp. Pl.: 518 (1753).

Only one variety is recorded for the island.

var. **tetrandra** (Humb. & Bonpl. ex Willd.) DC., Prodr. 2: 426 (1825). Fig. 2H.

A woody perennial herb or shrublet to 1 m; stems glabrous or almost so, with scattered patent prickles; stipules erect or spreading, lanceolate, acuminate, striate, setose on outer face and especially on the margins; leaves bipinnate, the pinnae in 1–2 pairs, the leaf rachis much shorter than the petiole in leaves with 2 pairs of pinnae so that the leaf appears ± palmate; pinnae multifoliolate; leaflets dark green to bluish green, tinged reddish, ± oblong, acute, asymmetric about the midvein, the upper face glabrous, the lower face sparsely appressed-pubescent to glabrous, the margins ciliate, sensitive; infloresences pedunculate, ellipsoid heads, these single or paired in the leaf axils or aggregated into a terminal raceme of heads; calyces yellowish green; corolla rosey-pink; stamen filaments white to pinkish; anthers creamish pink; fruit a craspedium of 2–3 articles, the articles glabrous, the replum setose.

Near base-camp, *Lewis* 1499 (fl. & fr.)!; estrada from base-camp to river, *Lewis* 1567 (fl. & fr.)!

Mimosa schrankioides Benth. in Hook. Lond. Journ. Bot. 5: 86 (1846).

Climber to 6 m; stems with a dense white, crisped pubescence and large patent golden-yellow setae intermixed, armed with numerous strongly recurved prickles; stipules linear-lanceolate, white pubescent and yellow setose; leaves bipinnate, the pinnae reduced to a single pair; petioles pale green, ± oblong to oblong-lanceolate, apiculate, ± symmetrical about the midvein except at base, both faces appressed-pubescent but hairs shorter and more sparse above; inflorescences pedunculate globose heads, single or paired in the axils of the upper leaves or aggregated into a terminal raceme of heads; corolla red to dull vermilion; stamen filaments very pale pink, almost white; fruit a craspedium of 3–4 articles, these densely pubescent with short white hairs and a few yellowish setae intermixed, the replum with patent yellow setae up to 3 mm long.

Scrubby boundary of savanna.

Estrada from base-camp to Santa Rosa, *Lewis* 1667 (fl. & fr.)!; near Casa de Santa Rosa, *Milliken* 371 (fl. & fr.)!

Mimosa sensitiva L., Sp. Pl.: 518 (1753). Fig. 2G.

Scrambling climber; stems ± ribbed, these together with the petioles and peduncles with a pubescence of short white hairs and deflexed yellow setae and sharp recurved prickles intermixed; stipules ovate-lanceolate, striate, pubescent and setose; leaves bipinnate, the pinnae reduced to a single pair; leaflets 3–4 per pinna with one basal leaflet much smaller than the other 2 or 3, dark green with a bluish tinge, strongly asymmetric about the midvein, ± falcate, the upper face glabrous, the lower face appressed-pubescent and appressed-setose, the margin also with appressed setae; inflorescences pedunculate, globose heads in the leaf axils or terminal as a raceme of heads; stamen filaments white; fruit a craspedium of 3–4 articles, these finely pubescent and setose, the replum setose.

Scrubby waste ground.

Near base-camp, *Lewis* 1500 (fl. & fr.)!

15. **ACACIA** Mill.

About 1200 spp., mainly subtropical to tropical, in Australia, Asia, Africa and the Americas. The taxonomy of neotropical species is not well known.

1. Flowers in spikes; fruits breaking up into 1-seeded, oblong articles ... ... **A. articulata**
1. Flowers in heads; fruits not breaking up into 1-seeded articles:
   2. Flower heads 5–6 mm across; leaflets of mature leaves under 5 × 1 mm ... ... **A. paniculata**
   2. Flower heads 1 cm or more across; leaflets of mature leaves over 5 × 1 mm:
      3. Robust climber (on Maracá) ... ... **A. riparia**
      3. Tree (on Maracá) ... ... **A. sp. aff. glomerosa**

Acacia articulata Ducke in Arch. Jard. Bot. Rio de Janeiro 3: 73 (1922). Fig. 2C.

Tree to 6 m with scandent branches, young stems sparsely armed with scattered, recurved prickles; leaves bipinnate, the pinnae opposite, an ellipsoid nectary near the base of the petiole, just above the pulvinus; leaflets yellowish-green, the blade asymmetric about the midvein; inflorescence a panicle of spikes; fruit yellowish green, maturing reddish brown, the surface finely and shortly pubescent with some intermixed minute reddish glands, 8–12-seeded, the seeds in ± indehiscent 1-seeded woody articles, these breaking up from the main body of the fruit as individual seed packages, the fruit margins slightly but regularly constricted between the articles and possibly partly remaining as a persistent replum (the fruit thus a true craspedium).

Riverine forest

Assuming that this plant has been correctly identified from the fruiting collection cited below then it was previously only recorded from the type locality near Monte Alegre in Pará, a considerable distance from the island. The two syntypes, *Ducke* 16038 and 16494 were collected from periodically inundated land by the river Gurupatuba. The species is distinguished from all other Brazilian Acacias by its articulate pod.

Rio Urariquera between base-camp and Maracá house, *Lewis* 1527 (fr.)!

Acacia sp. aff. glomerosa Benth. in Hook. Lond. Journ. Bot. 1: 521 (1842). "Faveca de paca."

Tree, DBH 35 cm; leaves bipinnate, the petiole bearing a dome-shaped gland;

leaflets obliquely lanceolate, strongly asymmetric about the main vein, generally 4-costate from the base, discolourous, darker green on upper face, sparsely appressed-pubescent below, glabrous above, the margins ± ciliate, flowers unknown; fruit a flat, coriaceous, glabrous, dehiscent pod, ca. 9–12-seeded.

Terra firme forest.

Closely related to *A. glomerosa* but in the absence of flowering material we are unable to identify the plant to species.

Ca. ½ km from base-camp on road to Santa Rosa, *Ratter et al.* 5488 (fr.)!

Acacia paniculata Willd., Sp. Pl. 4: 1074 (1806). Fig. 2E.

A robust climber (perhaps sometimes a tree) to 10 m; stems usually sparsely armed with scattered recurved prickles; leaves bipinnate, the pinnae opposite; petiole bearing an ellipsoid nectary, smaller disc-like nectaries between the upper pairs of pinnae; leaflets mid-green, under 5 × 1 mm, individual inflorescences globose, these grouped in fascicles which are compounded into a complex pseudopanicle; flowers fragrant; stamen filaments white; fruit a compressed, dehiscent pod, green tinged purplish when immature, 10–15-seeded; seeds ellipsoidal, dark olive brown, the fine pleurogram open at the upper end.

Riverine forest.

Margin of Rio Urariquera, *Lewis* 1501 (fr.)!; Furo de Maracá, Casa de Maracá, *Lewis* 1690 (fr.)!; margin of Furo de Maracá, *Henderson et al.* 562 (fl.)! (this recorded as a tree to 8 m); near landing [leading to base-camp], *Milliken* 346 (fl.)!

Acacia riparia Kunth, Nov. Gen. & Sp. 6: 276 (1824).

Robust climber to 25 m, the main stem ribbed, 8 cm in diameter, the wood very porous and light in weight, younger stems usually armed with scattered prickles which assist in the climbing habit; leaves bipinnate; individual inflorescences globose, these grouped in fascicles compounded into a complex pseudopanicle; calyx and corolla sparsely pubescent; stamen filaments creamish; fruit a compressed, dehiscent pod.

Terra firme forest.

A species requiring detailed study. The specimen cited below consists of fallen flowers and a small wood sample only.

Estrada from estação sede to Santa Rosa, *Lewis* 1696 (fl.)!

16. **INGA** Mill.

Since Bentham's monograph (1875) the only other attempt at a revision was by Pittier (1916). The most recent publication on the genus at a regional level is Poncy's treatment of the species of French Guyana in Mém. Mus. Nat. d'Hist. Nat. Paris, nouvelle série, 31: 1–124 (1985). The genus is in urgent need of a thorough taxonomic revision and until this is available many other collections remain impossible to name to species. This is particularly true of sterile ecological vouchers from Maracá which are here cited after treatment of the six taxa so far recorded for the island.

About 350 spp., tropical and subtropical America.

1. Leaf rachis prominently winged, the wing 5–10 mm wide; corolla densely tomentose; fruit terete, sometimes twisted or somewhat coiled, up to 35 × 1 × 1 cm, densely golden-brown tomentose; seeds embedded in a pulpy mesocarp ... ... **I. sp. aff. ingoides**
1. Leaf rachis not winged or if slightly so then wing under 5 mm wide and leaflets never more than 2 pairs per leaf; corolla and fruits without the above combination of characters:
  2. Leaf rachis and petiole terete, neither winged, margined or channelled; flowers pedicellate; racemes sometimes paired in leaf axils ... ... **I. sp. aff. cylindrica**
  2. Leaf rachis and/or petiole either narrowly winged or channelled ; flowers sessile; spikes fasciculate or aggregated into panicles:
    3. Fruit 1.5–1.8 cm wide, swollen over the seeds, the valves thus undulate and the seed number evident ... ... **I. alba**
    3. Fruit over 2 cm wide, the valves not undulate and the seed number not evident:
      4. Fruit coriaceous, laterally compressed, up to 5 cm long, under 2 mm in thickness, the apex emarginate with the style persisting as a small mucro, the valves prominently reticulate-veined at right angles to the margins ... ... **I. bourgonii**
      4. Fruit woody, over 10 cm long, over 5 mm in thickness, the apex rounded to obtuse, the valves rugose and/or pustular but lacking reticulate venation:
        5. Tree of riverine forest to 8 m tall; fruits rugose; leaflets drying leaden greyish ... ... **I. splendens**
        5. Tree of terra firme forest, 20–25 m tall; fruits pustular; leaflets drying olive-greenish brown ... ... **I. sp. aff. splendens**

Inga alba (Sw.) Willd., Sp. Pl. 4: 1013 (1806).
Mimosa alba Swartz, Fl. Ind. Occ. 2: 796 (1800).

Tree 20–30 m, DBH 20–35 cm; bark smoothish, orange-brown with transverse striations and horizontally stretched abcission scars; outer slash pinkish red, inner slash white, giving a slightly sweet odour; leaves pinnate with (2–)3–5 pairs of opposite leaflets; petiole and leaf rachis without a wing or rachis, sometimes slightly winged near apex, a gland between each leaflet pair, leaflets glabrous or almost so, acute to acuminate at apex; flowers in spikes, these clustered into axillary fascicles or compounded into short panicles arising from the older wood below the leaves; calyx shortly pubescent; corolla white, 2–4 times length of calyx; fruit 15 × 1.5–1.8 cm, glabrous to minutely puberulous or pulverulent, prominently transversely veined, swollen over the seeds and the valves thus undulate, the somewhat thickened margins occasionally constricted between the seeds, 5–10-seeded; mesocarp sweet, edible.

Terra firme forest.

Regeneration plots [near base-camp], *Milliken & Miller* 777 (fr.!); near end of the Trilha de Preguiça, transect tree 7/4, *Ratter & Milliken* 6129V (sterile)!

Inga bourgonii (Aubl.) DC., Prodr. 2: 434 (1825).
Mimosa bourgoni Aubl., Hist. Pl. Guian., Fr.: 941 (1775).

"Ingá xixica"

Tree, 5–20 m, DBH up to 25 cm; bark smoothish, orange-brown; outer slash pinkish-red, inner slash white; leaves pinnate, 2–3(–4) jugate, the rachis slightly winged especially towards the apex, a flat or slightly raised discoid gland between each pair of leaflets; leaflets coriaceous, discolourous, the lower face paler; flowers in short-pedunculate, compact spikes, these clustered into axillary fascicles, a persistent ovate-triangular bract ½ way up the peduncle; calyx moderately short-pubescent; flowers strongly scented with a pleasant fruit-like odour, the corolla white, glabrous but for a few hairs on the lobes, ca. 3 times length of calyx; fruit compressed, glabrous, transversely-veined, the margin somewhat thickened, up to 5 × 2.5 cm (but larger from specimens outside Maracá).

Riverine and terra firme forest.

Milliken & de Mota 608 is sterile except for some very young inflorescences but seems best placed here.

Ilha da Nova Olinda, *Milliken & Bowles* 440 (fl.!); near base-camp, *Milliken & de Mota* 608 (flower buds)!; trail one (by base-camp) *Ratter et al.* 6323 (fr.)!

Inga sp. aff. cylindrica (Vell.) Mart. in Flora 20, Beibl. 2: 114 (1837).

Tree, 6 m, 15 cm diam.; young stems, leaf and inflorescence rachises golden-yellow tomentose; leaves pinnate (2–)3 jugate, the rachis without a wing, a small discoid gland between each leaflet pair; leaflets narrowly elliptic, apically acute to acuminate, coraiceous, ± bullate, the upper face glabrous at maturity but for a fine pubescence on the midvein, the lower face sparsely pubescent; flowers scented, shortly-pedicellate, in short-pedunculate elongated axillary racemes, these sometimes 2 per axil; calyx and corolla shortly pubescent, the white corolla ± 3–4 times length of calyx; fruit unknown.

Riverside vegetation.

Near Cachoeira de Prumama [Pururamé], Furo de Santa Rosa, *Milliken et al.* 722 (fl.)!

Inga sp. aff. ingoides (Rich.) Willd., Sp. Pl. 4: 1012 (1806). Fig. 3D, Plate 4E.

Small to medium-sized, wide-crowned tree, 6–10 m, the branches often hanging down into the river; younger stems golden-brown tomentose; older branchlets with numerous whitish, pustular lenticels; leaves pinnate, the petiole usually without a wing but the rachis winged between each pair of leaflets, the wing 5–10 mm wide; discoid glands between each pair of leaflets, the leaflets membranaceous to subcoriaceous, the upper face mid-green, the lower face paler, sparsely to moderately pubescent on both faces; flowers in spikes, these solitary or clustered into axillary fascicles; calyces greenish, tomentose; corolla creamish or greenish yellow, pilose, ca. 2 times calyx length; stamen filaments creamish white; fruit indehiscent, terete, long cylindrical, sometimes twisted and/or coiled, up to ca. 35 × 1 × 1 cm, longitudinally deeply sulcate, densely golden-brown tomentose; seeds embedded in a sweet, edible, pulpy, white mesocarp.

River margins.

A very common species around the eastern end of the island, often forming a dense, almost impenetrable tangle along the river banks.

Between base-camp and Maracá House, *Lewis* 1433 (fl. & fr.)!; north side of island, Igarapé opposite entrance of Rio Trairão into Furo de Santa Rosa, *Lewis* 1550 (fl.); Furo de Maracá, *Henderson & Lima* 564 (fr.)!; extreme east of island [given as 'extreme west' but this must be wrong], *Hopkins et al.* 612 (fl.)!; Furo Pananá de Firmino of Rio Urariquera, *Hopkins et al.* 635 (fr.)!; same locality, *Hopkins et al.* 701 (fr.)!; Furo de Maracá near eastern end of island, *Sette Silva & Lima* 5818 (fl.)!

Inga splendens Willd., Sp. Pl. 4: 1017 (1806). Fig. 3K, Plate 4F.

Tree, 4–8m (–30 m in the Guianas); bark warty, greyish brown, mottled, slash pinkish; leaves pinnate, 1–2 jugate, the rachis winged near the apex, the wing 4–5 mm wide, a large, ± sessile, discoid gland between each leaflet pair; leaflets coriaceous, glabrous or almost so, discolourous, deep bottle green on upper face, yellowish green on lower face, broadly elliptic to obovate, the terminal pair considerably larger than the lower pair, up to 20 × 10 cm; flowers in long-pedunculate congested spikes, these solitary or clustered in the axils or aggregated into terminal panicles; bracteoles persistent, ovate-lanceolate, pubescent; calyx densely pubescent, corolla whitish; fruit thick, woody, glabrous, straight or ± falcate, indehiscent, up to 15 × 3.5 × 1 cm.

Riverine forest.

North side of island at Três Igarapés, *Lewis* 1563 (fr.)!; edge of Furo de Maracá, *Henderson et al.* 573 (fl.)!; Furo Pananá de Firmino of Rio Urariquera, *Hopkins et al.* 638 (fr.) (NY, photo K)!; same locality, *Hopkins et al.* 702 (fr.)!; Furo de Santa Rosa, *Hopkins et al.* 873 (fl.)!

Inga sp. aff. splendens Willd., Sp. Pl. 4: 1017 (1806).

"Kwaracuá" (in Maiongong dialect).

Tree 20–25 m, DBH 20–25 cm; first branching at 10 m from ground; young branchlets with pustular lenticels; bark dark brown or greyish white with large green lichen patches and numerous dot-dash lenticels; outer slash orangish, inner slash yellow or white; leaves pinnate, (2–)3 jugate, the rachis without a wing, a discoid gland between the leaflet pairs; leaflets coriaceous, glabrous, elliptic to rotund, acute to short acuminate at apex, up to 20 × 10 cm; flower unknown; fruit as for *I. splendens* but larger, up to 17 × 5 cm, the surface rough and pustular.

Terra firme forest and on steep slopes of open forest over weathered schist at 150-200 m altitude.

Evidently a species closely related to *I. splendens* but a larger tree of inland forest with bigger fruits and leaflets drying olive-greenish brown, not leaden grey. Flowering material is needed before making a final decision upon the correct placement of this taxon.

Western tip of island near Pururamé falls, *Lewis & Milliken* 1687 (fr.)!; south of the Cachoeria de Fumaça, *Milliken et al.* 503 (fr.)!

Milliken & Bowles 258, collected up-river from the island at Igarapé Aracacá is a specimen of *Inga sertulifera* DC., not recorded from the island but to be expected at the western tip.

Sterile collections unnamed to species.

Milliken et al. 461 and 477; *Ratter et al.* 5676V, 5757V, 5768V, 5877V, 6127V and 6135V.

17. **ABAREMA** Pittier

About 20 spp. in Central and South America.

Abarema jupunba (Willd.) Britton & Killip in Ann. N.Y. Acad. Sc. 35 (Mimos & Ceasalpin. Colomb.): 126 (1936).

Acacia jupunba Willd., Sp. Pl. 4: 1067 (1806); *Pithecolobium jupunba* (Willd.) Urb. (1900).

"Ingá do sapo".

Tree, 20–40 m, DBH 30–90 cm (the 40 m tree branching into two secondary trunks at 1.5 m, sometimes with low buttresses; bark flaking in chunks; outer slash orange, inner slash yellow; leaves bipinnate, the pinnae in (2–)3–4(–6) pairs, a disc or cushion-shaped gland between the lowest pinnae pair and often between some of the other pairs of pinnae also, similar but smaller glands also present just below the insertion of most of the leaflet pairs; leaflets in 2–5(–8) pairs, subcoriaceous, obliquely ovate or rhomboid, discolourous, the upper face dark bottle green, shiny, glabrous, the lower face paler, dull, inconspicuously pubescent on the blade, the midvein and margins more evidently hairy; tertiary venation prominent on both faces; inflorescences axillary with a rusty-brown tomentum; stamens well over 10 per flower, the filaments creamish pale yellow; fruit a circinnate or hellicoidal, flat pod, dehiscent along the lower margin, somewhat constricted between the seeds and swollen over the seed cavities, ca. 10 cm (uncoiled) × 0.8–1.6 cm, the valves red on the inner face, ca. 10-seeded; seeds whitish with a bluish green area within the pleurogram.

Terra firme forest.

Western tip of island near Pururamé falls, *Lewis & Milliken* 1688 (fl.)!; ca. 5 km NW of base-camp, *Milliken & de Mota* 827 (fl.)!; Trilha de Preguiça, ca. 5 km N of the Casa de Maracá, *Ratter et al.* 6309V (sterile)!

18. **ALBIZIA** Durazz.

Nielsen in Advances in Legume Systematics, part 1: 180 (1981).
About 150 spp., mainly tropical.

1. Fruit thick, woody and dehiscent, the sutures 5–6 mm wide, the mesocarp sticky and sweet, the valves fractured at right angles to the sutures; leaflets usually over 5 × 2.5 cm ... ... **P. nuriense**
1. Fruit not as above, either chartaceous or semi-coriaceous, dehiscent or not, sometimes breaking up into 1-seeded segments but not leaving a persistent replum; leaflets up to 4 × 2 cm:
  2. Pinnae in 1–5 pairs per leaf:
    3. Flowers pedicellate in umbels or corymbose-umbels; fruit with ± straight margins, breaking up into 1-seeded segments ... ... **A. corymbosa**
    3. Flowers sessile in heads aggregated into fascicles, these compounded into reduced panicles; fruit dimorphic, either ± moniliform and ± breaking up into 1-seeded segments or ± straight-margined and not breaking up ... ... **A. polyantha**
  2. Pinnae in well over 5 pairs per leaf:
    4. Stipitate pinhead-shaped glands between upper 5–8 pairs of leaflets; fruit dehiscent; large tree of terra firme forest with rough orangey brown bark ... ... **P. aff. elegans**
    4. Glands lacking between leaflets; fruit apparently indehiscent; probably ± breaking up into 1-seeded segments; medium-sized tree of riverine or seasonally flooded forest ... ... **A. glabripetala**

Albizia corymbosa (L.C. Rich.) G.P. Lewis & P.E. Owen comb. nov. Fig. 3J, Plate 4D.

Mimosa corymbosa L.C. Rich. in Act. Soc. Hist. Nat. Paris 1: 113 (1792); *Pithecolobium corymbosum* (L.C. Rich.) Benth. (1844); *Samanea corymbosa* (L.C. Rich.) Pittier (1925); *Arthrosamanea corymbosa* (L.C. Rich.) Kleinh. (1940).

Tree, 8–12(–30) m with a broad crown, the lower branches occasionally somewhat scandent, DBH 15–80 cm; bark brown or blotched greyish and brown, smooth; outer slash orange exuding a sticky orange sap, inner slash white; leaves bipinnate, the pinnae in 2–5 pairs, a small cushion-shaped gland between most pairs of pinnae and most pairs of leaflets; leaflets in (4–)8–12(–14) pairs, obliquely subrhombic to obovate-oblong, subcoriaceous, discolourous, the upper face deep bottle green, shiny, the lower face paler, dull, sparsely pubescent to glabrous, the margins occasionally ± ciliate; inflorescences axillary or terminal; flowers faintly scented, dimorphic, the central ones ± sessile and larger, the outer ones pedicellate, the filiform pedicels ca. 4–8 mm long; calyces sparsely appressed-pubescent; corollas dull red, ca. 2–2.5 times length of calyx, funnel-shaped, sparsely to moderately appressed-pubescent especially on the lobes; stamen filaments cream; style white, stigma pale green; fruit coriaceous, glabrous, ca. 8–10 × 1–1.8 cm, breaking up into 1-seeded articles without leaving a persistent replum, the articles acting as float-chambers and readily distributed by water; seeds oblong, olive-brown outside the brown pleurogram, greenish within it.

Riverine forest.

The species falls within Nielsen's 1981 concept of *Albizia* sensu lato and is characteristic of a small group of species with fruits disarticulating into 1-seeded segments. Irwin in Mem. N.Y. Bot. Gard. 15: 107 (1966), discussing the similarities between this species and *Pithecellobium gonggrijpii* Kleinh., draws attention to the apparent difference in flower colour of the two, as recorded on field notes, with *Albizia corymbosa* (*P. corymbosum)* apparently having a cream or yellowish corolla and greenish white stamen filaments and *P. gonggrijpii* with a pink corolla and white filaments. However, *Milliken* 548, evidently a specimen of *A. corymbosa,* records the corollas as dull red and the filaments as creamish, throwing some doubt on the accuracy or consistency of the observations compiled by Irwin.

Margin of the Furo de Maracá, *Lewis* 1505 (fr.)!; northern side of island at Três Igarapés, ca. 1 km from sub-camp, *Lewis* 1558 (sterile)!; Furo de Maracá, *Lewis* 1678 (fl.)!; south of Cachoeira de Fumaça, *Milliken et al.* 519 (sterile)!; same general locality, *Milliken* 548 (fl.)!

Albizia sp. = Pithecellobium aff. elegans Ducke in Arch. Jard. Bot. Rio de Janeiro 3: 64 (1922).

Tree, 30 m +, DBH 66 cm; bark orangey brown, rough; outer slash orange-brown, inner slash white; leaves bipinnate, the leaf and pinnae rachises densely pubescent with rusty brown, spreading hairs especially along the upper face, the petiole bearing a shiny cushion-shaped gland ca. 2 mm in diam.; pinnae in ca. 12–13 pairs; leaflets in ca. 12–30 pairs, membranaceous, oblong to subrhombic, inequilateral at the base, rounded at apex, the upper face sparsely pubescent with short hairs, the lower face pubescent with longer spreading or subappressed hairs; a stipitate, pinhead-shaped gland, ca. 1.5–2 mm long, between the upper 5–8 pairs of leaflets; fruit a dehiscent, subligneous pod with ± straight, thickened margins and undulate valves somewhat constricted between the seeds.

Terra firme forest.

Assuming the fruit, collected from the forest floor, truly belongs to the rest of *Milliken et al.* 650 then this plant is certainly closely related to *Pithecellobium elegans* and *P. pedicellare* (both belonging to *Albizia* sensu lato) and probably represents an undescribed species, but it is unknown in flower and only known by the one collection. The stipitate glands between the upper leaflet pairs are particularly diagnostic within the group of closely related taxa, although similar glands do occur in other species of *Albizia*.

E. of Cachoeria de Prumama [= Pururamé], Furo de Santa Rosa, *Milliken et al.* 650 (fr.)!

Albizia glabripetala (H.S. Irwin) G.P. Lewis & P.E Owen comb. nov.
Pithecellobium glabripetalum H.S. Irwin in Mem. N.Y. Bot. Gard. 15: 109 (1966).
"Angico".

Tree 10–12 m (–18 m in British Guiana), DBH 15–60 cm; first major branching at ca. 5 m; bark smoothish with some knobbly warts on main trunk, greyish; slash whitish yellow to yellow with a strong unpleasant odour; outer wood mottled, inner wood white; leaves bipinnate, pinnae alternate to subopposite; an ovate nectary on petiole just above pulvinus; leaf rachis channelled along upper surface, pubescent; leaflets discolourous, deep bluish green above, paler below, pubescent especially on upper face and main vein below, their blades asymmetric about the main vein; flowers in heads or umbels, these compounded into showy terminal panicles; fruit compressed, glabrous, 6–15-seeded, irregularly constricted between the seeds, and larger fruits often with almost straight margins, tending to break up into ± 1-seeded articles (thus a loment) and occasionally to leave a persistent replum (then technically a craspedium), seeds fawn-coloured, compressed, a small U-shaped pleurogram on each face.

Riverine forest; seasonally flooded forest; low seasonally flooded carrasco woodland.

Closely related to *Albizia niopioides* (Spruce ex Benth.) Burk., a species from Santarém, Pará and not known to occur on the island, and to *Albizia polycephala* (Benth.) Killip, a species from southern Brazil. The type, *Jenman* 364 was collected in British Guiana and other collections come from that country and from eastern Venezuela.

Down-river from base-camp towards Furo de Santa Rosa, *Lewis* 1590 (sterile)!; on picada 'Ae', *Milliken* 88 (fr.)!; N of Casa de Maracá, *Milliken et al.* 116 (fr.)!; without exact locality, *Ratter et al.* 5622V (sterile)!; on triha de Preguiça, 2–3 km N of the Casa de Maracá, *Ratter et al.* 6263 (fr.)!

Albizia polyantha (Spreng.f.) G.P. Lewis in Legumes of Bahia: 164 (1987). Fig. 3 G & H, Plate 4 A & B.
Acacia polyantha Spreng.f. in Spreng., Syst. Veg. 5: 3 (1828); *Acacia multiflora* Spreng. (1826), non Kunth (1824); *Pithecellobium multiflorum* (Kunth) Benth., pro parte (1844); *Arthrosamanea polyantha* (Spreng.f.) Burkart (1949).

Medium to large tree, 8–15 m, sometimes multiple-trunked, usually with an umbrella crown with the lower branches reaching to within a metre or two of the ground, DBH (10–)40–80 cm; bark fairly smooth, lenticelled, mottled various shades of grey, pale brown and orangish green (much of this mottled colouring being due to lichen growth), large chunks of bark occasionally flake off trunk, slash creamish yellow to yellowish orange, odourless with no resin or sap exuded; leaves bipinnate; leaflets discolourous, dark green above, paler below, their blades asymmetric about the main vein; petiole and leaf rachis channelled along upper surface with an ellipsoid nectary restricted within the channel on the lower ¼ of the petiole and small round nectaries between the upper one or two pinnae pairs; inflorescences axillary or terminal, reduced panicles consisting of fascicles of heads widely spaced along the main rachis; buds green; flowers greenish yellow, delicately scented with a subtle lemon odour; stamen filaments creamish; fruit a loment, variable in form, irregularly constricted between the seeds and 12–13 mm wide or regularly constricted and ± 8 mm wide, maturing purplish black, glabrous, ca. 10–15-seeded; seeds olive brown.

Riverine forest on white sand.

An apparently variable species, particularly in the form of the fruit which although essentially a loment does show a tendency to break-up to leave a persitent replum and thus be a true craspedium.

Between base-camp and Maracá house on the Furo de Maracá, *Lewis* 1438 (fr. — narrower form)!; up-river from Maracá house, *Lewis* 1533 (fr. — wider form)!; up-river from Maracá house, *Lewis* 1534 (fr. — narrower form)!; as for Lewis 1533 (same tree), *Lewis* 1652 (fl.)!

Albizia sp. = Pithecellobium nuriense H.S. Irwin in Mem. N.Y. Bot. Gard. 15: 107 (1966) ('*Pithecellobium nuriensis*'). Fig. 3C.

"Coronha".

Large tree 30–40 m; DBH 80–90 cm; bark pale fawn, flaking in large chunks; slash white, inner slash yellow; leaves bipinnate; leaflets subcoriaceous, sparsely to densely pubescent on lower face, especially on the venation, on upper face pubescent on veins only; fruits thick, woody, indehiscent, the sutures dilated, 5–6 mm wide, the valves finely pubescent, fractured at right angles to the sutures, the mesocarp very sticky and sweet.

Tall terra firme forest.

The two collections cited below consist of fruits and leaflets collected from the forest floor. The lack of adequate herbarium material from the island, together with confused taxonomy and the resulting complicated nomenclature of the *Albizia saman — Albizia inopinata* 'complex' make positive placement of the Maracá species difficult. There seems however, little doubt that *Pithecellobium nuriense* H.S. Irwin is the same as the Maracá plant but Barneby (pers. comm. 1987) suggests that this is conspecific with *Albizia inopinata* (Harms) G.P. Lewis. That combination was made by G.P.L. in Legumes of Bahia (1987) in order to place several Bahian collections including *Mori & King* 12191, the fruit of which, together with that of the type, *Ducke* 15248, is twice the width of Maracá material although otherwise very similar. The Maracá fruits are more akin in size to those of true *Albizia saman* but in the latter the fruits are glabrous (including those of the type from Caracas) and without the right-angled fracturing on the valve surfaces. Barneby again (pers. comm. 1987) states, however, that in much of SW Amazonia the fruits of *Albizia saman* sens. lat. are pubescent like those of *A. inopinata*. Based on available information, Maracá material could be placed in either this pubescent-fruited group of *A. saman* or as a narrow-fruited form of *A. inopinata* but is most 'comfortable' in *Pithecellobium nuriense* for which there is no available combination in *Albizia*. The group evidently requires further detailed study and at present we prefer not to recognise *A. inopinata* and *P. nuriense* as conspecific. Fruits are sold in Boa Vista market (Ratter 5723) and the vernacular name of "coronha" or gun-stock might indicate that the wood has been used for this purpose.

Estrada from base-camp to Santa Rosa, *Lewis* 1422 (fr.)!; same locality, *Ratter et al.* 5723 (fr.)!

19. **ENTEROLOBIUM** Mart.

Being revised by A.L. Mesquita who, with M.F. da Silva, described a new species *E. barnebianum* in Acta Amazonica 14 (1/2): 153 (1984).

About 6 to 8 spp. in the W. Indies, Central and South America.

Pinnae in 5–11 pairs; leaflets in 12–30 pairs, each over 10 mm long × ca. 3 mm wide, corolla 4–6 mm long, 2–3 × length of calyx; the calyx white pubescent especially on the lobes; fruits ca. 8–12 cm in diam. ... ... **E. cyclocarpum**
Pinnae in 11–24 pairs; leaflets in 40–80 pairs, each under 10 × 1.5 mm; corolla 3–4 mm long, only the lobes exserted beyond the calyx; the calyx densely rusty brown tomentose; fruits smaller ... ... **E. schomburgkii**

Enterolobium cyclocarpum (Jacq.) Griseb., Fl. Brit. W. Ind.: 226 (1860). Fig. 3A.
Mimosa cyclocarpa Jacq., Fragm. Bot.: 30 & Tab. 34, fig. 1 (1801).

Very large tree to 50 m, DBH 1.50–1.90 m, crown spreading, lateral roots spreading out 10 m on surface of forest floor; bark slash orange-red with slight sweet smell; branchlets glabrous but with raised pustular lenticels; leaves bipinnate, the pinnae in ca. 5–11 opposite pairs, an elliptic, black nectary ca. ⅓ of the way up the petiole and ± sunken nectaries between the upper (3–)4 pinnae pairs; leaflets in 12–30 pairs, narrowly oblong, acute at apex, obliquely rounded at base, very discolourous, drying leaden grey on upper face, much paler on lower face, the blade very asymmetric about the midvein, sparsely pubescent, glabrescent below, glabrous above; flowers in pedunculate heads, these often compounded into axillary fascicles or ramiflorous pseudopanicles, strongly scented; calyces tubular, the short teeth densely pubescent, the tube less so; corolla pale greenish, moderately pubescent; stamen filaments white or creamish; anthers pale yellow; fruit reddish brown to black, indehiscent, laterally compressed except over the seed cavities, curved in a complete circle or part spiral, ca. 8–12 cm in diam., the margin undulate; seeds dark brown with a pale pleurogram, compressed ovoid-ellipsoid, embedded in a sweet tasting edible mesocarp, rattling loose as the fruit dries.

Terra firme forest.

Milliken & Lima 30 report that spider monkeys eat the fruit.

Forest trails close to base-camp, *Hopkins et al.* 906 (fr.)!; same general locality, *Milliken & Lima* 30 (fl. & fr.)!

Enterolobium schomburgkii (Benth.) Benth. in Trans. Linn. Soc. Lond. 30: 599 (1875) & in Mart., Fl. Bras. 15 (2): 457 (1876). Fig. 3B.
Pithecellobium schomburgkii Benth. in Hook., Lond. Journ. Bot. 3: 219 (1844).

Tree 20–30(–40) m, DBH 30–100 cm; crown flattened; bark yellow, pale brown or greyish, rough, flaking; slash of outer wood pinkish red, orangish or yellowish brown, granular, inner wood white or yellow, slash exudes a colourless sticky sap; leaves bipinnate, the pinnae in ca. 11–24 opposite to subopposite pairs, a dome-shaped or elliptic, black nectary just above the pulvinus on the petiole and similar but smaller nectaries between the upper pinnae pairs; leaflets in 40–80 pairs, small, generally under 5 × 1.5 mm, strongly asymmetric about the midvein, discolourous, darker green, shiny and glabrous above, dull and pubescent below; flowers in pedunculate heads clustered into axillary fascicles; calyx and

Fig. 3., all × ⅔. **A**: *Enterolobium cyclocarpum* (from Hopkins et al. 906); **B**: *Enterolobium schomburgkii* (from Lewis 1419); **C**: *Albizia sp.* = *Pithecellobium nuriense* (see text) (from Lewis 1422); **D**: *Inga sp. aff. ingoides* (from Hopkins et al. 635); **E**: *Zygia glomerata* (from Lewis 1529); **F**: *Zygia divaricata* (from Lewis 1554); **G**: *Albizia polyantha*, one fruit form (from Lewis 1533); **H**: *Albizia polyantha*, second, distinct fruit form (from Lewis 1438); **J**: *Albizia corymbosa* (from Lewis 1505); **K**: *Inga splendens* (from Hopkins et al. 702).

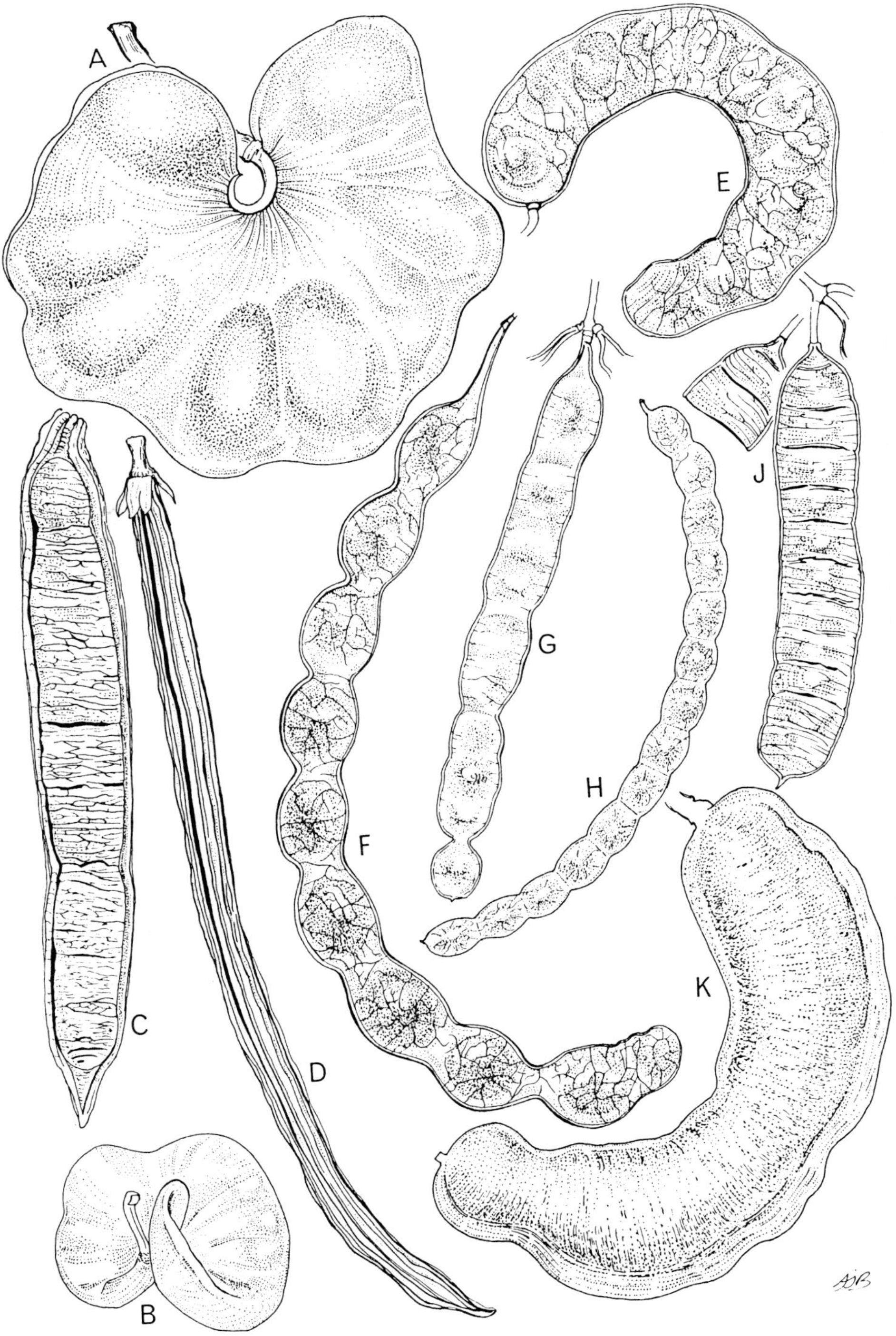
A
B
C
D
E
F
G
H
J
K
AJB

corolla densely golden tomentose; stamen filaments white, usually drying brick-red; fruit reddish brown, ± a smaller version of *E. cyclocarpum*; seeds yellowish ochre with a darker areole.

Terra firme forest.

Estrada from base-camp to Santa Rosa, *Lewis* 1419 (fr.)!; E of Cachoeria de Pururamé, Furo de Santa Rosa, *Milliken et al.* 640 (sterile)!; E of Cachoeira de Prumama [= Pururamé], Furo de Santa Rosa, *Milliken et al.* 668 (fr.)!; ca. ½ km from Santa Rosa station, *Ratter et al.* 5421 (sterile)!; Proctor gap plot, *Ratter et al.* 5754V (sterile)!; *Lewis* 1699 (seedlings only) probably belongs here.

20. **ZYGIA** P. Br.

About 20 spp., Central and tropical South America.

Key to be used as a provisional guide to identification pending Rico Arce's revision of the genus.

1. Secondary venation of leaflets with 2–4 curved veins running up ⅔ – ¾ blade length ... ... **Z. latifolia**
1. Secondary venation not as above:
   2. Petiole and pinnae rachises densely pubescent with spreading hairs ... ... **Z. cauliflora**
   2. Petiole and pinnae rachises glabrous ... ... } **Z. divaricata** / **Z. glomerata** (see notes in text)

Zygia cauliflora (Willd.) Killip in Trop. Woods 63: 6 (1940).
Inga cauliflora Willd., Sp. Pl. 4: 1021 (1806); *Pithecolobium cauliflorum* (Willd.) Mart. (1837).

Shrub or tree, 4 (–20) m, 4 m specimen with main stem 4 cm diam.; bark pale, slightly peeling; leaves bipinnate, the pinnae reduced to a single pair each with (3–)5 leaflets, the terminal leaflets ca. 2 × the size of the basal one, the petiole and pinna rachis densely pubescent with spreading hairs, these obscuring a disc-shaped gland at the apex of the petiole between the pinnae, a similar, smaller gland between the terminal pair of leaflets; leaflets ovate to obliquely elliptic, the base inequilateral, the apex acute to acuminate, subcoriaceous, glabrous but for a few hairs on the veins; inflorescences cauliflorous, sometimes several clustered in the axils of the older branchlets, the flowers sessile on short-pedunculate heads; the peduncles (3–)4–8 mm long; calyces ca. 1 mm long, ciliate on the margins; corolla whitish, funnel-shaped, several times longer than calyx, pubescent especially on the lobes; the staminal tube exserted from the corolla, the filaments mauve towards their free apices; fruit straight or curved, flat, coriaceous, ca. 5–15(–20) × 1.5 cm, constricted at irregular intervals, dehiscent along one margin.

Low forest bordering savanna.

E tip of island, *Milliken & de Mota* 607 (fl.)!

Zygia divaricata (Benth.) Pittier in Trab. Mus. Com. Venez. 2: 69 (1927). Fig. 3F.
Pithecolobium divaricatum Benth. in Hook. Lond. Journ. Bot. 3: 213 (1844), based on Bongard's manuscript name *Inga divaricata.*

Multiple-trunked tree to 10 m; bark greyish; leaves bipinnate with a single pair of pinnae, each pinna with 3 leaflets, the basal one smaller than the terminal pair, a disc-shaped gland between the pinnae and a similar but smaller gland between the terminal leaflets; leaflets mid-green, ovate to narrowly elliptic, subcoriaceous, glabrous, inequilateral at base, acuminate at apex; inflorescence a few-flowered, short-pedunculate head, ramiflorous; flowers sessile; corolla white; staminal tube exserted, the filaments reddish pink apically; fruit a yellowish-green, arcuate, dehiscent pod, opening along the lower margin, coriaceous, glabrous, regularly or irregularly constricted betwen the seeds, the suture ± thickened, the valves sometimes contorted, ca. 8–10-seeded.

Terra firme forest at margin of igarapé.

North side of island at Três Igarapés, ca. 2 km along picada from sub-camp, *Lewis* 1554 (fr.)!

Zygia glomerata (DC.) Pittier in Trab. Mus. Com. Venez. 2: 70 (1927). Fig. 3E, Plate 4C.

Inga glomerata DC., Prodr. 2: 438 (1825); *Pithecolobium glomeratum* (DC.) Benth. (1844).

"Ingá branca".

Shrub to small tree, 3–5 m, multiple-branched from base; bark greyish or brownish yellow, rough, vertically fissured; leaves bipinnate, almost sessile, with a single pair of pinnae, each with 2–3 glabrous, narrowly to broadly elliptic coriaceous leaflets, a disc-shaped gland between the pinnae and a smaller gland between the terminal pair of leaflets; inflorescence a sessile or short-pedunculate, few-flowered head, cauliflorous, the heads sometimes culstered on short, woody shoots on the older branchlets; fruit a yellowish green dehiscent pod, coriaceous, glabrous, arcuate, the margins somewhat thickened, slightly but irregularly constricted between the seeds, ca. 10–15-seeded.

Riverine vegetation.

This species is very closely related to *Z. divaricata* and may ultimately prove to be conspecific.

Between base-camp and Maracá House, *Lewis* 1441 (fr.)!; Furo de Maracá, ca. 3 km up-river from Casa de Maracá, *Lewis* 1529 (fr.)!; Furo Pananá de Firmino on S side of island, *Hopkins et al.* 751 (fr.)!; near Casa da Maracá, *Milliken et al.* 174 (fr.)!; Santa Rosa area, ca. 4 km NE of Station buildings, *Ratter et al.* 6155 (fr.)!

Zygia latifolia (L.) Fawc. & Rendle, Fl. Jam. 4, pt. 2: 150 (1920).

Mimosa latifolia L., Syst. Nat., ed. 10: 1310 (1759); *Pithecolobium latifolium* (L.) Benth. (1844).

Tree, 7(–12) m; bark smooth, mottled grey and white; leaves bipinnate, almost sessile, pinnae 1 pair, a disc-shaped gland between the pinnae and a smaller, similar gland between the terminal pair of leaflets; leaflets 3–5 per pinna, mid-green, darkening with age, elliptic, acute to rounded at apex, subcoriaceous, glabrous but for a few hairs on midvein and secondary veins, the secondary venation characteristically with 2–4 curved veins running up ca. ⅔– ¾ of blade length; inflorescences cauliflorous, few-flowered, subsessile, or shortly pedunculate heads densely clustered on the older branchlets; corolla funnel-shaped, striate, the staminal tube exserted, with the filaments red apically; fruit a curved, ligneous, dehiscent pod, the margins somewhat thickened and irregularly constricted between the seeds.

Riverine forest; seasonally flooded várzea forest.

North side of island at Três Igarapés, ca. 1 km up igarapé from sub-camp, *Lewis* 1562 (fr.)!; near Casa de Maracá, *Milliken et al.* 112V (sterile)!

21. **SWARTZIA** Schreber

Cowan in Fl. Neotropica 1: 1–228 (1968) (revision of the neotropical species). About 140 spp., nearly all neotropical.

1. Leaves 1–3-foliolate, the terminal leaflet ca. 2 × the others in size; buds 3–4 mm in diam.; stamens isomorphic (i.e. of similar form but about half functionally fertile, the others staminoidal); gynoecium glabrous; fruit glabrous, 1-seeded ... ...**S. arborescens**
1. Leaves 5–13-foliolate, the petiole never winged, the rhachis sometimes obscurely winged or ridged between the upper leaflet pairs; buds 5–12 mm in diam.; stamens heteromorphic, the larger ones few in number, the smaller ones many; gynoecium sericeous; fruit densely tomentose to sparcely pubescent, glabrescent, 1–8-seeded:
  2. Flowers bipistillate; tall tree to 30 m of terra firme forest; large stamens 2–3; fruit 1-seeded, drupe-like, obovate in outline ... ... **S. dipetala**
  2. Flowers with a single pistil; tree to ca. 10 m, essentially of savanna or forest margins; larger stamens 4–25; fruit 1–7-seeded (usually ± moniliform with 3–7 seeds):
    3. Leaflets moderately to densely pubescent on lower face; buds 8–12 mm wide; larger stamens 20–25 in number; fruit rugose, densely yellowish orange tomentose ... ... **S. grandifolia**
    3. Leaflets glabrous or almost so on both faces; buds (5–)6(–9) mm wide; larger stamens 4 in number; fruit minutely golden pubescent, glabrescent ... ... **S. laurifolia**

Swartzia arborescens (Aubl.) Pittier in Journ. Wash Acad. 11: 157 (1921).
Possira arborescens Aublet, Pl. Guian. 2: 934 (1775).

Tree to 15 m, DBH 20 cm; bark greyish brown, slightly rough; wood off-white; stipules subulate, ca. 4–6 mm long, ± persistent; leaves 1–3-foliolate, the petiole and rachis winged; stipels present but minute; leaflets ovate to lanceolate, the terminal one usually elliptic, the apices acute with retuse tips, coriaceous, glabrous but for the midvein, the tertiary venation finely reticulate and prominent; inflorescence an axillary, few-flowered raceme; buds globose, 3–4 mm in diam.; calyx segments 3–4, at length deciduous, glabrous; the single petal yellow, glabrous; stamens isomorphic, the functional ones ca. 10–15, + ca. 15 staminoidia; gynoecium glabrous; fruit glabrous, elliptic in outline, laterally compressed, stipitate, ca. 3.5–4 × 1.5–2 cm, at length dehiscent, 1-seeded.

Terra firme forest near river.

Furo de Santa Rosa, *Milliken & de Mota* 573 (fl. & fr.)!

Swartzia dipetala Willd. ex Vog. in Linnaea 11: 173 (March 1837).
Swartzia dicarpa Moricand ex Meisner (Aug. 1837); *Swartzia microstylis* Benth. (1840).

"Mututi de terra firme".

Tall tree to 30 m, DBH 70 cm; bark pale brown, flaking; outer slash pale brown, inner slash white; sap red; stipules linear-lanceolate, caducous; leaves pinnate with (2–)3–4 pairs of leaflets plus a single terminal one; stipels present; leaflets chartaceous to coriaceous, strongly discolourous, often drying blackish above and much paler below, ovate to ovate-lanceolate or elliptic, the apex acute, glabrous but for a few hairs on the midvein, the tertiary venation finely reticulate; inflorescence axillary, racemose or compounded into a panicle of racemes; buds globose to elliptic in outline, 5–6 mm in diam.; calyx lobes usually 4, glabrous on inner face, densely tomentose on outer face, in consequence the two faces strongly discolourous on drying; the single (rarely 2 ± equal) petal(s) yellow, globose; larger stamens 2 or 3, the smaller stamens many; gynoecium sericeous, bipistillate; fruit yellow, minutely strigulose, 2.5–3.5 × 1.5–2 cm, obovate in outline, stipitate, the aril covering ca. ½ the seed.

Terra firme forest.

S of Cachoeria de Fumaça, *Milliken et al.* 455 (sterile)!; Proctor's 6th plot [near base-camp], tree no 0599, *Ratter et al.* 5696 (flowers from forest floor)!; road to Santa Rosa, ca. ½ km from base-camp, *Ratter et al.* 5422V (sterile)!

Swartzia grandifolia Bong. ex Benth. in Hook. Journ. Bot. 2: 85 (1840). Fig. 4B, Plate 5A.

Shrub to medium-sized tree, 2–10 m, DBH (2.5–)12–18 cm; bark greyish brown, smooth, slightly fissured; young branchlets, petioles, peduncles, leaf and inflorescence rachises with a dense rusty brown tomentum; stipules lanceolate to ovate, up to 10 × 3 mm, tomentose, caducous; leaves pinnate with (3–)5–6 opposite pairs of leaflets plus a single terminal one; stipels sometimes present but obscure and not to be confused with the free tips of the rachis wings; leaflets mid-green, chartaceous or subcoriaceous, ovate to oblong or oblong-elliptic, the apex obtuse to rounded or acute, the upper face sparsely pubescent, glabrescent, occasionally ± nitid, the lower face pubescent, the venation on both faces pubescent; leaf rachis narrowly winged (sometimes conspicuously so) at least between the upper pairs of leaflets, the wing widening towards the base of each leaflet pair and then forming short, free, usually ± incurved tips; inflorescence a ramiflorous or cauliflorous raceme, the broadly ovate bracts persistent; buds globose to oval in outline, 8–12 mm in diam.; calyx segments 4, reflexing as ovary develops, thick, woody, yellowish velutinous on outer face, glabrous on inner face, often persisting to mature fruit; single petal deep lemon yellow with purplish veins, sericeous on outer face especially on the veins, glabrous on inner face; stamens dimorphic, the larger ones ca. 20–25, the smaller ones many; ovary sericeous with a 5–12 mm long gynophore; fruit a yellowish orange, tomentose, ± moniliform, rugose, stipitate, dehiscent pod, (1–)5–8-seeded; seeds egg-shaped, fleshy, the testa thin, arillate with an orange edible aril surrounding ⅔ of the seed circumference.

Savanna and forest margins; low campina forest on white sand; occasionally in riverine forest and seasonally flooded várzea forest.

Base-camp environs, *Lewis* 1443 (fr.)!; western tip of island, parana from Furo de Maracá to Furo de Santa Rosa, *Lewis & Milliken* 1682 (fl.)!; near base-camp, *Hopkins et al.* 568 (fl)!; trail behind Casa de Santa Rosa, *Hopkins et al.* 601 (fl.)!; Santa Rosa area, *Lima & Nelson* 735 (fl.)!; near Casa de Maracá, *Milliken et al.* 104V (sterile)!; near base-camp, *Ratter et al.* 5375 (fr.)!; without detailed locality, *Ratter et al.* 5635V (sterile)!; Santa Rosa, *Ratter et al.* 5687 (fl.)!; same locality, *Ratter et al.* 5699V (sterile)!; same locality, *Ratter et al.* 5708V (sterile)!; Proctor gap plot [near base-camp], *Ratter et al.* 5747V (sterile)!; near Igarapé camp, Fumaça trail, *Ratter et al.* 5863V (sterile)!; Santa Rosa area, ca. 4 km NE of Station buildings, *Ratter et al.* 6158A (fl.); same locality, *Ratter et al.* 6179 (fr.)!

Swartzia laurifolia Benth. in Hook., Journ. Bot. 2: 87 (1840). Fig. 4C, Plate 5B.

Shrub to small tree to 3–10 m, DBH of 8 m tree 10 cm; bark mottled greyish brown, smooth, not fissured; copious red sap exuded from the bark slash and cut branches; stipules falcate-elliptic, acute, up to ca. 25 × 7 mm, often persistent; leaves pinnate with 2–5 pairs of opposite leaflets plus a single terminal one;

minute stipels at each leaflet pair; leaf rachis sometimes with a slightly expanded margin but never conspicuously winged; leaflets dark bottle-green, coriaceous, glabrous or almost so and ± nitid on upper face, very sparsely pubescent, glabrescent below, elliptic to oblanceolate-elliptic, the apex bluntly acute; inflorescence racemose, usually in ramiflorous fascicles; bracts ± persistent; buds ± globose, (5–)6(–9) mm in diam.; calyx segments 4, reflexed and persistent with the fruit, minutely tomentose on outer face, glabrous on inner face; single petal buttercup or pale canary yellow, glabrous but for a few hairs on the venation of the outer face; larger stamens ± 4, smaller stamens many, filaments yellow, anthers orangish brown; gynoecium sericeous; fruit a ± moniliform, ± rugose, stipitate, dehiscent pod, the valves minutely golden-pubescent, glabrescent, 1–7-seeded; seeds egg-shaped with a thin greyish yellow testa and orange fleshy aril, pendulous from fruit after dehiscence.

Open forest and savanna, on sandy soil.

Estrada from Santa Rosa to base-camp, near Casa de Santa Rosa, *Lewis* 1448 (fl. & fr.)!; trail behind Casa de Santa Rosa, ca. 1 km from river, *Hopkins et al.* 598 (fl)!; same locality, *Hopkins et al.* 875 (fr.)!; same general locality, *Nelson & Ferraz* 1528 (fr.)!; same locality, *Ratter et al.* 5598 (fl.)!; same locality, *Ratter et al.* 5695 (fl.)!; Santa Rosa area, ca. 4 km NE of Station buildings (JP 8388), *Ratter et al.* 6148 (sterile)!; same locality (JP 8389), *Ratter et al.* 6149 (sterile)!

Ratter et al. 5422V, a sterile collection from near Santa Rosa, is, as yet, unnamed to species.

22. **ZOLLERNIA** Wied.-Neuw. & Nees

de Mello Filho & de Andrade in Atas Simpos. Biota Amazonica 4 (Bot.): 153–166 (1967); Yakovlev in Bot. Zhurn. 61(9): 1304–1307 (1976) (review of the genus).

Perhaps 12 or more species, some undescribed, mostly in Brazil and the Guianas but also in Central America.

Zollernia grandifolia Schery in Fieldiana, Bot., 28, No. 2. : 270 (1952).
"Pau Santo".

Tree, 20–ca. 35 m, DBH 15–45 cm, sometimes with small buttresses; bark chocolate-brown to black, peeling or scaling in large irregular flakes 'giving a shaggy appearance rather like the surface of an *Azteca* ant nest' (*Ratter & Milliken* 6246); wood white; stipules ovate to narrowly lanceolate, acuminate, held erect, glabrous, to ca. 10 mm long, persistent; leaves simple, the petiole wrinkled, channelled along the upper edge, the blade narrowly elliptic to oblanceolate, acuminate, subcoriaceous, glabrous, distinctly toothed along the margins with the marginal vein occasionally extending as a sharp prickle; inflorescence a diffuse panicle in the upper leaf axils; petals 5, yellow, oblong to suborbiculate; stamens 10, free; ovary ca. 5-ovulate; fruit drupaceous, indehiscent, glabrous, 1(?–2)-seeded.

Terra firme forest.

The four collections, cited below, were identified after comparison with Schery's original description of the species but there is no authoritatively named material at Kew for comparison.

S of Cachoeira de Fumaça *Milliken et al.* 471 (sterile)!; N end of trilha de Preguiça, very close to the Furo de Santa Rosa, *Ratter & Milliken* 6246 (fr.)!; Transecto de tatu, ca. 4 km from base-camp, *Ratter et al.* 5512 (sterile)!; near Igarapé camp, Fumaça trail, transect tree 2: 1/1, *Ratter et al.* 5869V (sterile)!

23. **ACOSMIUM** Schott

Yakovlev in Not. Roy. Bot. Gard. Edinb. 29: 347–355 (1969) excl. syn. *Dicraeopetalum.*
16 spp., S. Mexico to N. Argentina, mostly in Brazil.

Fruits woody, drying black, fissured at right angles to the sutures, shortly-stipitate, the stipe 1–3 mm long; lower surface of leaflets with a prominent tertiary venation, moderately pubescent ... ... **A. nitens**
Fruits membranaceous, drying brownish orange, the valves not fissured but with a reticulate venation, long-stipitate, the stipe 12–15 mm long; lower surface of leaflets with tertiary venation only slightly prominent, sparsely pubescent ... ... **A. tomentellum**

Acosmium nitens (Vog.) Yakovl. in Not. Roy. Bot. Gard. Edinb. 29: 353 (1969). Fig. 4A.
Leptolobium nitens Vog. in Linnaea 11: 394 (1837); *Sweetia nitens* (Vog.) Benth. (1865).

Tree to 4 m, trunk 10–50 cm diam., bark flaky, fissured, greyish-black; leaves pinnate; leaflets leathery, discolourous, bottle green and shiny above, whitish-green and hairy below, the terminal leaflet distinctly larger than the others; inflorescence racemose, flowers yellow, sweetly scented; fruit black, woody, 1–2-seeded, fissured at right angles to the suture.

Riverine forest.

A common species on white sandy banks of tributaries of the Rio Branco near Boa Vista (see *Lewis* 1609, a specimen in fruit).

Margin of the Furo de Maracá, *Henderson et al.* 593 (fl.)! (NY, photo. K).

Acosmium tomentellum (Mohl.) Yakovl. in Not. Roy. Bot. Gard. Edinb. 29: 352 (1969). Fig. 4G.
Sweetia tomentella Mohl. in Webbia 17: 242 (1963).
"Sicupira amarela".

Tree 5–12 m, trunk 20–35 cm diam.; bark greyish, often breaking off in large chunks, slash yellow quickly turning blackish-purple; leaves pinnate, 3–5-foliolate; leaflets opposite, discolourous, shiny above, dull below, semi-coriaceous, sparsely pubescent to glabrescent on the lower face; petiole, leaf rachis and petiolules sparsely pubescent to tomentellous; inflorescence a terminal panicle, young buds greenish, tomentose, bracteolate; flowers pleasantly scented, petals creamish to yellow; anthers orange-brown; fruit membranaceous, reticulate-veined, indehiscent, long stipitate, 1–2-seeded, wind dispersed; seeds brownish orange.

Riverine forest; on white sand amongst large boulders; low campina forest.

A little known species with no previous collections at Kew for comparison. The type, *Kuhlmann* 3239 was collected near Boa Vista, along the Rio Branco.

Furo de Maracá, *Lewis* 1654 (sterile with bark sample)!; Furo de Maracá, Corredeira dos Filhotes, *Lewis & Milliken* 1679 (fr.)!; Rio Urariquera, a 6 km baixando o rio, *Lima et al.* 817 (buds)!; Furo Paraná do Firmino, *Milliken & Bowles* 378 (fl.)!; on trilha de Preguiça, 2–3 km N of the Casa de Maracá, *Ratter et al.* 6241V (sterile)!

24. **ORMOSIA** Jacks.

Rudd in Contribs. U.S. Nat. Herb. 32 (5): 278–384 (1965) (the American species).

About 100 spp. in the tropics of S. America and SE Asia to NE Australia.

1. Both faces of leaflets glabrous; flowers yellow, the standard sometimes with a central red blotch; seeds black ... ... **O. flava**
1. At least the lower face of leaflets with some form of indumentum; flowers purplish; seeds red or bicoloured red and black:
   2. Tertiary venation of leaflets prominently raised on lower face, secondary venation looping just inside leaflet margin, each secondary usually linking up with the next; fruits densely velutinous at maturity; seeds scarlet; stems deeply sulcate and densely velutinous ... ... } **O. coarctata** / **O. sp. A**
   2. Tertiary venation of leaflets obscure or at least never as prominent as the secondaries, these straight and parallel and running to the leaflet margin, there curving to fade out parallel with the margin and not linking up with each other; fruit glabrous at maturity; seeds bicoloured red and black; stems neither deeply sulcate nor velutinous:
      3. Flowers 9–10 mm long; tree of terra firme forest on west of island; leaflets acute ... ... **O. paraensis**
      3. Flowers 13–15 mm long; tree of riverine forest on east of island; leaflets acuminate ... ... **O. smithii**

Ormosia coarctata Jacks. in Trans. Linn. Soc. 10: 363 (1811). Fig. 4F.

Tree to 18 m, DBH 30 cm; bark peeling in fine flakes; young stems reddish or yellowish velvety tomentose; stipules linear-deltoid, ca. 3–5 mm long; leaves pinnate, 5–11-foliolate; leaflets opposite, oblong-obovate to elliptic, the apex acute to shortly acuminate, the upper face glabrous, the lower face moderately to densely pubescent, the midvein and secondary veins very prominent below, somewhat sunken above; inflorescence a ? terminal panicle; flowers 12–15 mm long; calyx shortly tomentose; petals dark purple; fruit dehiscent, subligneous, up to 5 × 3 cm, densely tomentose with brownish hairs, 1(–3)-seeded; seeds bright red, hanging from open fruits by a thin funicle.

Terra firme forest.

Rudd, page 361 (1965), describes the seeds of this species as 'bicoloured red and black' but Stirton, working on neotropical species of the genus, has identified the collection cited below, with its unicoloured red seeds and larger than normal fruits, as *O. coarctata*.

Forest trails close to base-camp, *Hopkins et al.* 668 (fr.)!

Ormosia flava (Ducke) Rudd in Contribs. U.S. Nat. Herb. 32(5): 298 (1965).
Clathrotropis? flava Ducke in Arch. Jard. Bot. Rio de Janeiro 3: 134 (1922), *Ormosiopsis flava* (Ducke) Ducke (1925).
"Cernambi do indio", "Tentoarana".

Tree to ca. 30 m, DBH 80 cm, with small low buttresses; bark yellow-fawn, scaling in long, narrow, irregular flakes which roll up at the edges; young stems with appressed reddish brown hairs, glabrescent; leaves pinnate, 5–11-foliolate; leaflets opposite to subopposite, subcoriaceous, elliptic to oblong-elliptic, acute to acuminate at apex, the blade glabrous, midvein below somewhat pubescent, secondary veins inconspicuous; flowers 15–18 mm long; petals yellow, the standard sometimes with a central red blotch on inner face; fruit a dehiscent woody pod, glabrous at maturity, ca. 3 × 1.5–2 cm, 1(–3)-seeded; seeds black, shiny, globose.

Terra firme forest.

Trail 1 [near base-camp], *Ratter et al.* 6325 (fr.)!

Ormosia paraensis Ducke in Arch. Jard. Bot. Rio de Janeiro 4: 62 (1925).
Ormosia crassicarpa Pierce ex Pittier (1944).

Tree to 25(–40 off island) m, DBH 29 cm; bark pale, rough; outer slash yellowish brown, inner slash white; leaves pinnate 7–9(–15)-foliolate, leaflets opposite, coriaceous, elliptic-oblong, obtuse to acute at apex, the upper face glabrous, subnitid, the lower face minutely appressed-pubescent, dull, glabrescent, the midvein prominent below, sunken above, the secondary veins slightly raised below; flowers 9–10 mm long; petals purplish marked white; fruit a dehiscent woody pod, glabrous at maturity, 1–3-seeded; seeds bicoloured red and black.

Terra firme forest.

E of Cachoeira Pururamé, Furo de Santa Rosa, *Milliken et al.* 646 (sterile)!

Ormosia smithii Rudd in Contribs. U.S. Nat. Herb. 32(5): 339 (1965). Fig. 4E, Plate 6A.

Tree (4–)8–15 m, DBH of 15 m tree 50 cm, sometimes multiple-trunked; crown wide and spreading; bark mottled shades of grey, ± smooth; cut branchlets exude copious blood-red watery sap; young branchlets woolly pubescent; leaves pinnate, 5–9-foliolate; leaflets opposite, coriaceous or subcoriaceous, ovate to obovate-oblong, the apices usually with acuminate drip-tips, discolourous, the upper face dark bottle-green or deep bluish green, glabrous or with a few hairs on the veins, the lower face paler, sparsely to moderately short pubescent or ± farinose, the midvein woolly, prominent, the secondary veins pubescent, ± parallel to the leaflet margin and then looping to almost join up with each other; flowers 13–15 mm long; petals violet or purplish, the standard with a white blotch on inner face; fruit a dehiscent woody, glabrous, subnitid pod, 1–3-seeded, the margins somewhat constricted on fruits with more than 1 seed; seeds bicoloured red or orangish and black (ca. 10–25% of the testa surface black), shiny, pendent by funicle attachment after fruit dehiscence.

Riverine forest.

Rio Uraricoera, ca. 1.5 km down-river from landing stage leading up to base-camp, *Lewis* 1565 (fr.)!; Furo de Maracá, *Lewis* 1675 (fr.)!; Furo de Maracá, west of Corredeira dos Filhotes, *Lewis & Milliken* 1680 (fr.)!; Furo de Maracá, *Henderson et al.* 613 (fr.)! (NY); extreme east [given as 'west'] end of island, *Hopkins et al.* 614 (fr.)!

Ormosia sp. A
"Tento"

Tree, DBH 19 cm; stems and leaf-rachis densely golden yellow-brown tomentose; stipules linear, ca. 8 mm long, persistent; leaves pinnate; leaflets ovate to oblong-elliptic, acuminate, upper face glabrous but for a few hairs on the sunken midvein, lower face pubescent with the prominent midvein and secondaries tomentose; fruit a dehiscent woody, densely tomentose, 1–2-seeded pod; seeds scarlet, subglobose.

Terra firme forest.

Stirton has seen the specimen cited below and confirms that it represents a new species which he is writing up.

Near Igarapé camp, Fumaça Trail, *Ratter et al.* 5879 (fr.)!

Ratter et al. 5626 is a sterile collection of *Ormosia* unamed to species.

25. **ALEXA** Moq.

Ramirez, thesis published by the Central University of Venezuela, Caracas: 1–67 (1983) (Revision of the Venezuelan species).
About 10 spp. in Venezuela, the Guianas and Amazonian Brazil.

Alexa canaracunensis Pittier in Bol. Soc. Venez. Cienc. Nat. 7, 51: 308 (1942). Fig. 4H, Plate 5C & D.
"Tunadí" (in Maiongong dialect).

Tree 6–12(–30 in Venezuela) m; trunk 10–20 cm diam.; bark greyish with a green tinge, reticulately fissured into small, irregular, wavy-edged, ± vertically rectangular blocks; outer slash orangish with yellow horizontal markings, inner slash white, slash smells of freshly cut cucumbers, sap exuded from wound turns gelatinous after several days and forms a jelly-like coating over the cut area; phyllotaxy spiral (leaves in whorls), leaves pinnate; very young foliage mauvish purple; leaflets subopposite to opposite, elliptic with long drip tips, discolourous, dark green above, paler below; tiny round nectaries between some leaflet pairs; young stems subtending inflorescences hollow, their inner surface a dark chocolate brown colour; inflorescences erect, terminal, 10–15-flowered; flower buds greenish overlain by a dark brown, almost black, velvety pubescence, flowers at anthesis 6–8 cm long, petals creamish tinged green, in open flowers these rolled back and flower appearing pseudocaesalpinioid; stamens free; a large fleshy pear-shaped bract subtending each flower, these covered in up to 20 tiny round nectaries; fruit a dehiscent pod ± 15 cm long, velvety with a dark chocolate-brown indumentum, the thin green style often persisting to maturity, the valves twist at dehiscence to expose a white pithy endocarp; 2–5(–6) ellipsoid seeds embedded in the pith, these white ripening dark brown to black.

Lower slopes of terra firme forest; open forest on hill slopes and summits at ca. 150–200 m altitude; on 1 m deep orange clayey soil resulting from weathered schist and overtopped by a deep leaf litter; understorey rich in *Marantaceae*, *Rubiaceae*, *Musaceae* and *Heliconiaceae*; also on 75 cm brown earth above 25 cm black, eroded rock on hill summits.

A species worthy of the considerable attention it is currently receiving. C.H. Stirton, in his work on the neotropical species of tribe Sophoreae, has remarked upon the close affinity of the genus *Alexa* to the monotypic Australian legume *Castanospermum australe* (pers. comm. 1987). His taxonomic observations are supported by the recent findings of R.J. Nash and co-workers who have isolated the plant alkaloid castanospermine (first extracted from the seeds of *C. australe*) from leaves, pods and seeds of seven species of *Alexa*, including *A. canaracunensis* (Phytochemistry 27(5): 1403–1404 [1988]). Collaborative studies between the Jodrell Laboratory at the Royal Botanic Gardens, Kew and St. Mary's Hospital Medical School, London have demonstrated that castanospermine exhibits acitivity against the AIDS virus *in vitro* ("The Lancet": 1025, 31 Oct. 1987). Apart from the obvious advantages to man and his medicine these results are of great significance to the work of the conservation lobby in arguing against the continued destruction of the world's natural resources. The Maiongong Indians working on Maracá were unable to report any local uses of *Alexa canaracunensis*, so here we have a plant apparently unused by the indigenous population but perhaps eventually proving to be of major importance to mankind.

The species is one of the dominant leguminous trees in the forests on the west of the island. Of equal abundance, growing in exactly the same habitat is the closely related *Clathrotropis macrocarpa*, the cut bark of which also has a cucumber odour and produces a jelly-like exudate. Although abundant in overall number, the individuals of *A. canaracunensis* are scattered throughout the forest and each tree has only one or two inflorescences in flower at any one time. Flowering is sequential from base to apex of each inflorescence and a 15-flowered inflorescence will only have one or two flowers at anthesis. The fruit crop of any one tree is unlikely to be greater than 3 at any given time.

The hollow stems of the young branchlets and inflorescences house small black ants of the genus *Procryptocerus* (specimens identified by B. Bolton of the British Museum of Natural History, 8.III.1988) and their larvae. These ants feed on secretions from the small nectaries on the bracts and foliage and in consequence can be found running about the tree from leaf to leaf and flower to flower and presumably act, at least in part, as a defense against visiting herbivores. Milliken observed "flowers defended by large red ants feeding from glands at the flower base" (*Milliken et al.* 622) and seedlings grown by R. Miller at base-camp developed foliage nectaries on the first leaf and these were visited by small red ants.

The type, *Cardona* 420 was collected in Guyana at Canaracuni on the headwaters of the Rio Caura.

Western tip of island near Pururamé falls, *Lewis & Milliken* 1689 (fr. & seedlings)!; east of Cachoeira de Pururamé, Furo de Santa Rosa, *Milliken et al.* 622 (fl. & fr.)!; same general locality but at Pururamé encampment off island, *Lewis & Milliken* 1683 (fl., fr. & seedlings)!

26. **CLATHROTROPIS** Harms

6 spp. in tropical South America, principally the Amazon and Orinoco basins.

Clathrotropis macrocarpa Ducke in Tropical Woods 31: 16 (1932). Fig. 4D, Plate 5E & F.

"Kahadi" (in Maiongong dialect).

Tree to 20 m, DBH 20–40 cm, first branching of trunk at 6 m, crown open; bark rough, mottled with greyish and greenish lichens, vertically fissured with horizontally stretched lenticels; outer slash orange, inner slash white, bark slash smelling of freshly cut cucumber with a slight hint of lemons; leaves pinnate; leaflets opposite, coriaceous, ovate to narrowly ovate, held ± erect by pulvinules, discolourous, deep bottle green and shiny on upper face, paler whitish green on lower face; inflorescence an axillary or terminal, large, showy, tomentose panicle; flowers ca. 10–15 mm long, petals purplish; fruit a thick, woody, dehiscent pod, ca. 12.5–15 × 6–8 cm, the lower suture 7–8 mm wide, the upper suture much wider with two woody submarginal ridges, the valves golden-yellow tomentose, fracturing at dehiscence into a jigsaw pattern and the exocarp breaking away from the endocarp, 1–2-seeded; seeds 'overgrown', fleshy, up to ca. 8 × 5 cm, on drying the thin testa separating from the cotyledons and turning orangish brown, furrowed parallel to the margin.

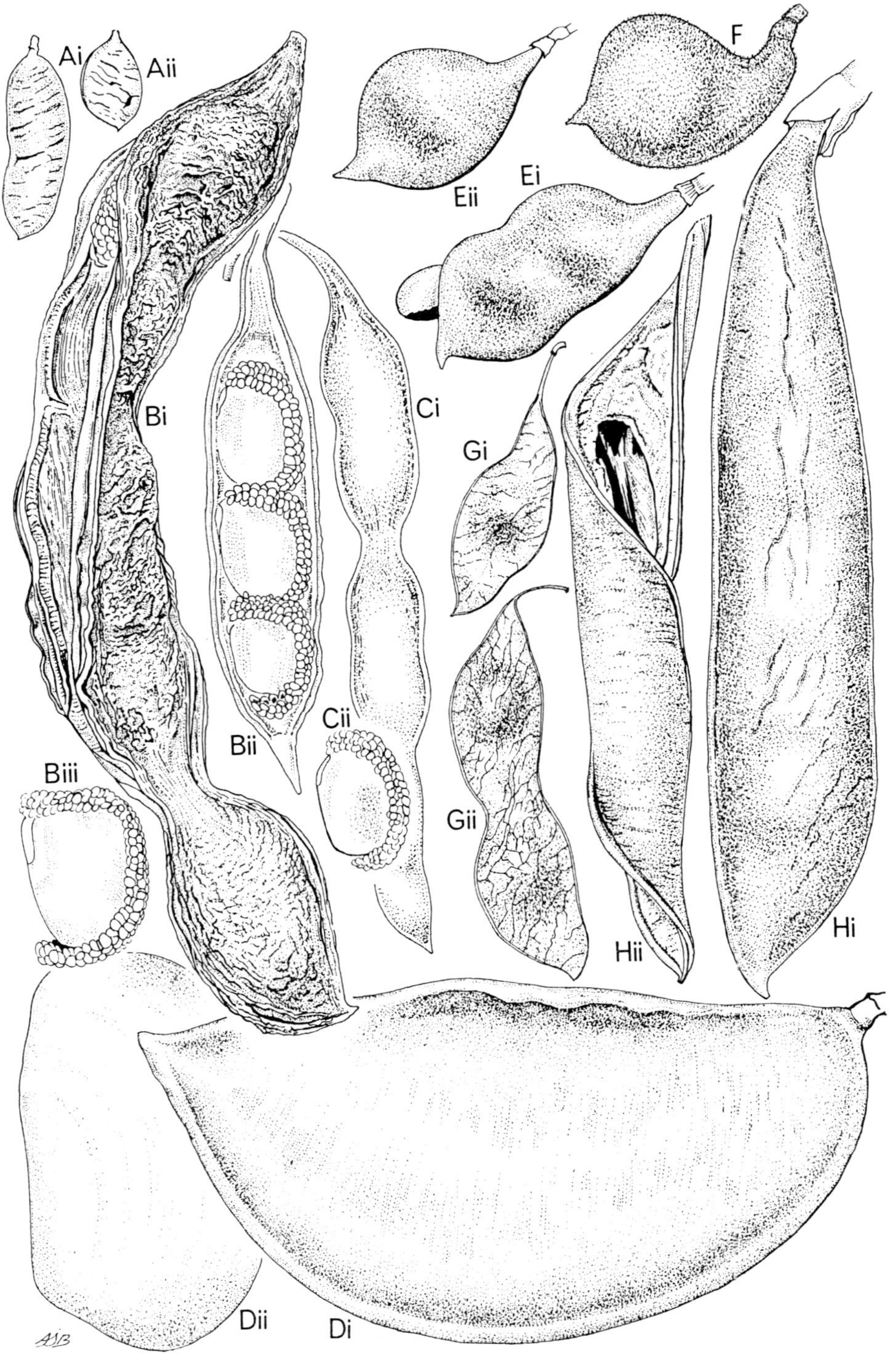
Ai
Aii
Bi
Bii
Biii
Ci
Cii
Di
Dii
Ei
Eii
F
Gi
Gii
Hi
Hii
AJB

Open terra firme forest on slopes of small hill at western tip of island. On orange clayey soil resulting from weathered schist between 150–200 m altitude.

One of the dominant tree species at the western tip of the island near the Pururamé falls. As with *Alexa canaracunensis*, the other dominant legume in these forests, the cut bark smells of cucumbers and old slash wounds are covered in a gelatinous secretion. The trees reach the upper canopy of the forest of 20 metres and in October 1987 produced a heavy fruit crop. According to the Maiongong Indians, crushed bark is used as a remedy to the painful sting of ponerine ants ("tucandeira").

Western tip of island near Pururamé falls, *Lewis & Milliken* 1684 (fr.)!; same locality, *Milliken et al.* 630 (fr.)!

27. **ANDIRA** A.L. Juss.

N.F. Mattos in Acta Amazonica 9(2): 241–266 (1979) (revision of the Brazilian species).
30–35 spp. in tropical America, one extending to Africa.

Leaves with 9–11(–15) chartaceous to subcoriaceous leaflets; flowers (10–)12–15 mm long ... ... **A. inermis**
Leaves with 7–(9–11) coriaceous leaflets; flowers 15–18 mm long ... ... **A. surinamensis**

Andira inermis (Wright) Kunth, Nov. Gen. & Sp. 6: 385 (1824).
Geoffroea inermis Wright in Lond. Medic. Journ. 8: 256 (1787).

Tree, (8–20)–30 m, DBH ± 90 cm, with a spreading crown; bark fairly smooth, somewhat flaking, grey to chocolate brown; wood slash orangey brown; leaves pinnate, 9–11(–15)-foliolate; leaflets subcoriaceous, glabrous on upper face, finely appressed pubescent on lower face; inflorescence a dense, well-branched axillary or terminal panicle, shortly brownish hairy; flowers pinkish to purplish red, ca. (10–)12–15 mm long; fruit an ovoid to globose drupe, sometimes with a broad, blunt longitudinal rib on one side, glabrous, 1–2-seeded, supported on a short thick stipe, pericarp fleshy, endocarp woody, when cut smelling of broad beans (*Vicia faba*); seeds ellipsoid.

Isolated patch of forest amongst savanna.

The only species, of an essentially neotropical genus, also native in parts of Africa; the type is from Jamaica but there is apparently no preserved specimen. A species common in swamps and marsh forests along the coastal plain of the Guianas, with fruits quite probably dispersed by bats. Milliken reports that many fruits on the ground had been gnawed.

Just below base-camp, *Milliken* 800 (fr.)!

Fig. 4., all × ⅔. **A**: *Acosmium nitens* (from environs of Boa Vista, Lewis 1609); **B**: *Swartzia grandifolia*, **i**, whole fruit, **ii**, packing of seeds in fruit, **iii**, seed (from Lewis 1443); **C**: *Swartzia laurifolia*, **i**, fruit (from Nelson 1528), **ii**, seed (from Hopkins et al. 875); **D**: *Clathrotropis macrocarpa*, **i**, fruit, **ii**, seed (from Lewis & Milliken 1684); **E**: *Ormosia smithii* (from Lewis 1675); **F**: *Ormosia coarctata* (from Hopkins et al. 668); **G**: *Acosmium tomentellum* (from Lewis 1679); **H**: *Alexa canaracunensis*, **i**, before dehiscence (from Lewis & Milliken 1689), **ii**, single valve after dehiscence (from Lewis & Milliken 1683).

Andira surinamensis (Bondt.) Splitg. ex Pulle in Enum. Plant Surinam.: 229 (1906). Fig. 5H, Plate 7A.

Geoffroea surinamensis Bondt., De cortice Geoffr. Surinam. 13 (1788); *Geoffroea retusa* Poir. (1808); *Andira retusa* (Poir.) Kunth (1824).

"Sowó" (in Maiongong dialect).

Tree (4–)5–12(–35 in the Guianas) m, DBH of 12 m tree 50 cm; bark dark brown with grey patches, flakey with some large chunks breaking off; slash salmon pink to orangey brown, no sap exuded, odourless; leaves pinnate, 7–(9–11)-foliolate; leaflets opposite, dark green, coriaceous, glabrous on upper face, finely appressed pubescent on lower face; inflorescence a lax, many-branched axillary or terminal panicle, tomentose; calyces dark purple to almost black, 5–6 mm long; flowers 15–18 mm long, petals pale pink, the standard with a central white blotch surrounded deeper pink on the inner face, the basal half of the blotch flecked red; ovary red, glabrous; fruit an ovoid, oblong or globose drupe, 1–2(–3)-seeded, pericarp fleshy, endocarp woody; seeds very hard, light brown with paler speckles.

Riverine forest, river-smoothed rocks amongst white sand; seasonally flooded forest; low campina forest.

Cut fruits smell strongly of broad beans (*Vicia faba*). Leaflets often have scattered, globose or columnar insect galls.

Furo de Maracá, Lewis *1671* (fl.)!; same locality, *Milliken et al.* 435 (fr.)!; between Igarapé Traira and W point of island, *Milliken & Bowles* 307 (fr.)!; north of Casa de Maracá, *Milliken et al.* 143V (sterile)!; without detailed locality, *Ratter et al.* 5624 (sterile)!; Santa Rosa, *Ratter et al.* 5721 (fr.)!

28. **HYMENOLOBIUM** Benth.

de Lima in Acta Amazonica 12(1): 41–48 (1982) & in Bradea 3(45): 404–405 (1983).

10–15 spp. in tropical South America.

Hymenolobium petraeum Ducke in Arch. Bot. Jard. Rio de Janeiro 1: 36 (1915). Fig. 5K, Plate 6B & C.

"Angelim de mata".

Very large emergent tree, 30–35(–50) m, DBH 1–1.40 m, crown open; bark blackish, flaking off in large woody chunks to reveal whorled patterning on trunk; outer slash pinkish with yellowish banding, inner slash creamish; sap clear; cut bark smelling of broad beans (*Vicia faba*) or garden peas (*Pisum sativum*); young branchlets exude a blood-red exudate when cut; young foliage and all parts of young inflorescence very finely appressed-pubescent; leaves pinnate; leaflets lime green tinged pink on the margins when young, oblong, retuse at apex; inflorescence a terminal panicle; calyx brownish purple; a pair of tiny persistent bracteoles just below calyx base; standard petal magenta with a central greenish yellow blotch on the inner face, wings and keel pinkish magenta with yellowish claws; ovary bright green; fruiting when leafless, the fruits purplish liver-coloured, membranaceous, indehiscent, glabrous, stipitate, slightly twisted at base, 8–9 × 2–2.5 cm, prominently reticulate-veined with one longitudinal vein 3–5 mm from the upper margin and sometimes a similar but less evident one close to the lower margin, the style persistent at apex, 1(–2)-seeded.

Terra firme forest.

A rare tree giving a spectacular show in flower and particularly in fruit — producing an "island" of bright colour above the upper forest canopy. Monkeys congregate in the crown and may eat the fruits. *Milliken* 53 was collected using a 10 bore shotgun!

Estrada from base-camp to Santa Rosa, *Lewis* 1698 (fr.)!; same locality, *Milliken* 53 (fl.)!; same locality, *Ratter et al.* 5457V (sterile)!

29. **MACHAERIUM** Pers.

Hoehne, Flora Brasilica 25(3): No. 128: 1–99 (1941) (includes a key to species). Neotropical species are currently being worked on by V.E. Rudd.

About 120 spp. from Mexico to Argentina (mostly in Brazil), one extending to W coasts of Africa.

1. Fruits woody, kidney-shaped, lacking a wing, glabrescent, water dispersed; foliage tending to dry blackish; inflorescence a densely congested axillary raceme or short panicle, much shorter than leaf that subtends it; leaves 3–5-foliolate ... ... **M. inundatum**
1. Fruits papery to subcoriaceous, winged, densely pubescent to glabrous, essentially wind dispersed; foliage drying greenish to greyish or brownish; inflorescence a large showy, usually terminal, panicle, usually larger than the leaves:
  2. Bracteoles enclosing calyx, especially in bud, ± persistent; wing and keel petals greenish, standard white or pinkish, the inner face flushed purplish centrally; fruit glabrous but for the occasional glandular-based hair, these especially prominent on the stipe of developing ovary ... ... **M. kegelii**
  2. Bracteoles $\frac{1}{3}$–$\frac{1}{2}$ length of calyx, persistent or caducous; petals mauvish, the standard with a central yellowish or greenish blotch on inner face; fruit moderately to densely tomentose; glandular-based hairs lacking:
    3. Secondary and tertiary venation strongly prominent on lower face of leaflets; leaflet apex acute, mucronate **M. quinata** var. **parviflorum**
    3. Only the midvein prominent on lower face of leaflets; leaflet apex rounded to obtuse, shallowly emarginate:
      4. Leaves with more than 10 leaflets, their margins plane, the secondary venation closely parallel ... ... **M. biovulatum**
      4. Leaves with less than 10 leaflets, their margins revolute ... ... **M. sp. A**

Machaerium biovulatum Micheli in Mém. Soc. Phys. Hist. Nat. Genève 34: 265 (1903).
Machaerium lilacinum Ducke (1922).

Tree, 6–20 m, DBH 5–24 cm, or woody vine flowering at 10 m; bark of tree with prominent warty lenticels and the occasional 1 cm long spine; outer slash brown, inner slash white; sap red; young stem with a brown felt-like indumentum; stipules subligneous, triangular to lanceolate, acute to acuminate, ± persistent; leaves pinnate; leaflets alternate, oblong to oblanceolate, obtuse to emarginate and mucronate at apex, increasing in size towards leaf apex, upper face glabrous, lower face sparsely to moderately appressed-pubescent especially on the midvein, the secondary venation very evident, ± parallel and closely spaced, scallariform, joining the marginal vein and branching only slightly; inflorescence an axillary or terminal panicle, nearly all parts with a dense velvety tomentum; bracteoles ovate, striate, persistent; petals pale mauve, the standard with a central yellow blotch on its inner face, the outer face densely silvery pubescent; fruit a samara, drying a straw-yellow colour, finely tomentose, 1-seeded, slightly warty over seed-cavity, deeply concave between the wing and seed-cavity on lower margin.

Riverine forest and low, partially flooded forest inland.

Milliken et al. 724 is a woody vine and his 467 and 469 are both medium-sized trees. This variation in habit is confirmed by *Ratter et al.* 6276V who states that 'this species starts life as a tree but later becomes a liana'. The habit may also depend on type of habitat, as indicated by the collections cited below, the riverine collection being a vine and the inland forest collections being trees.

About 3 km S of Cachoeira de Fumaça, *Milliken et al.* 467 (sterile but including a bark sample)!; same locality, *Milliken et al.* 469 (sterile)!; near Cachoeira de Prumama [= Pururamé], *Milliken et al.* 724 (fl.)!; transect tree No. 48, *Ratter et al.* 6276V (sterile)!

Machaerium inundatum (Mart. ex Benth.) Ducke in Arch. Jard. Bot. Rio de Janeiro 4: 311 (1925). Fig. 5D, Plate 6F.
Drepanocarpus inundatus Mart. ex Benth., Comm. Leg. Gen.: 32 (1837).

Robust liana to 6 m, main stem ca. 3 cm diam., occasionally a multiple-stemmed shrub or small tree to ca. 3 m; bark greyish white, flaking in papery scales, terminal and some axillary branchlets modified into twining tendrils and these armed with pairs of recurved hooks which assist the climbing habit; sap blood-red, copious; stipules early deciduous; leaves pinnate, the petiole, rachis and leaflet petiolules drying black; leaflets alternate, broadly ovate, acute to acuminate at apex, usually cordate at base, deep bottle-green drying black or greyish green, coriaceous, glabrous, shiny; inflorescence an axillary or ramniflorous contracted panicle of spikes, all parts including the outer face of the standard with a dense golden-brown tomentum; fruit indehiscent, reniform to suborbicular, thick and woody, ca. 2.5–3 × 1.5–2 × 0.5 cm, pale greenish yellow with darker veins, clothed in a very fine golden indumentum, glabrescent with age to expose an exocarp that dries black, 1-seeded.

Riverside vegetation and tree-gaps in terra firme forest (as a liana) and in the scrubby marginal area of a damp campo (as a shrub).

Três Igarapés sub-camp on north side of island, ca. 2 km along picada from sub-camp, *Lewis* 1552 (fr.)!; down-river from base-camp towards E tip of island, *Lewis* 1594 (sterile)!; same locality, *Lewis* 1598 (fr.)!; Casa de Maracá, *Lewis* 1691 (fl.)! extreme east of island (given as 'west' on field label), *Hopkins et al.* 619 (fr.)!; Furo de Santa Rosa, close to sub-camp 'Igarapé Pedra Sentada', *Milliken* 11 (fr.)!; Furo de Maracá, *Milliken et al.* 760 (fl. & immat. fr.)!; Santa Rosa, *Ratter et al.* 5564V (sterile)!; Santa Rosa area, ca. 4 km NE of Station buildings, *Ratter et al.* 6157 (fr.)!

Machaerium kegelii Meisn. in Linnaea 21: 257 (1848). Fig. 5B.
Machaerium bracteatum Benth. (1860); *Machaerium steinbachianum* Hoehne (1939).

Large woody climber to 10 m; bark mottled greyish and white; sap oxidises blood-red; stipules subligneous, acuminate, ca. 1–1.5 cm long, persistent; leaves pinnate; leaflets subopposite to alternate, dark green, coriaceous, glabrous, elliptic to oblong, oblanceolate or obovate, acuminate (drip-tipped), the prominent tertiary venation reticulate; inflorescence a large, showy terminal panicle, all parts purplish brown tomentose; bracteoles ovate, persistent, enclosing the calyx in bud; wings and keel petals greenish, the standard whitish to very pale pinkish, flushed purple centrally on the inner face, brown tomentose on outer face; fruit a stipitate samara up to ca. 10 cm long, 1-seeded, the seed-cavity with glandular-based hairs especially near the stipe on young developing ovaries.

Terra firme forest at edge of igarapé.

North side of island at Três Igarapés, *Lewis* 1564 (fr.)!

Machaerium quinata (Aubl.) Sandw. in Kew Bull. 1931: 359 (1931). Fig. 5E.

Only the smaller-flowered variety is known from the island.

var. **parviflorum** (Benth.) Rudd in Phytologia 24: 121 (1972).
Machaerium ferrugineum (Willd.) Pers. var. *parviflorum* Benth. in Mart., Fl. Bras. 15 (1): 253 (1862); *Machaerium nervosum* Vog. (1837).

Woody liana to 8 m with a lobed trunk; main stem 15 cm diam.; slash yellowy white; leaves pinnate; leaflets alternate, elliptic to oblong or oblanceolate, apically acute, mucronate, coriaceous, discolourous, the upper face glabrous, the lower face with a dense felty tomentum and a prominent reticulate tertiary venation; inflorescence paniculate, all parts with a golden-brown felty tomentum; bracteoles much shorter than calyx; petals lilac, the standard with a greenish-white central blotch on inner face; fruit an olive-yellow-brown, 1-seeded, stipitate samara with a velvety indumentum.

River bank vegetation; terra firme forest.

Proctor plot 1 [near base-camp], *Milliken* 80 (fl. & fr.)!; just inland of Furo de Santa Rosa, *Ratter et al.* 6196 (fr.)!; Furo de Maracá close to E tip of island, *Ratter et al.* 6297 (fr.)!

Machaerium sp. A Fig. 5C.

Large woody climber, the main stem attaining 30 cm diam. and armed with pairs of shiny, recurved thorns; stipules woody, ca. 3–5 mm long, ± persistent; leaves pinnate; leaflets alternate, elliptic, oblong or oblanceolate, semi-coriaceous, the upper face dark green, glabrous, the lower face paler with a fine appressed pubescence; inflorescence a large terminal panicle, all parts with a fine golden-brown tomentum; fruit a 1-seeded samara very similar to that of *M. biovulatum*.

Riverside vegetation.

V.E. Rudd (pers. comm., 1988) states that the collection, cited below, represents a new species in Section *Reticulata* and that its "best match is with a similar, unidentified collection, *Rosa & Cordeiro* 1505, from the same general area". She will be writing up the new species. *Lewis* 1588 and *Ratter et al.* 5741V, both sterile collections, may well belong here.

Between base-camp and Maracá house, *Lewis* 1439 (fr.)!

Milliken et al. 118V is a sterile collection unnamed to species.

30. **DALBERGIA** L.f.

Hoehne, Flora Brasílica 25(3) No. 126: 3–33 (1941) (includes a key to Brazilian species). Brazilian species currently being revised by A.M. de Carvalho at the University of Reading, England.

About 100 spp., pantropical.

Dalbergia riedelii (Benth.) Sandw. in Kew Bull. 1931: 358 (1931). Fig. 5F.
Ecastaphyllum monetaria Pers. var. *riedeli* Benth. in Journ. Linn. Soc. 4, Suppl.: 50 (1860); *Hecastophyllum riedeli* (Benth.) Radlk. in Koepff, Uber anat. char. Dalberg.: 41 (1892).

Robust woody liana to 8 m or more, main stem 1.5 cm diam., smooth, dark brown, occasionally with tendrils; leaves pinnate; leaflets alternate, yellowish green with a rusty brown tomentum when young, discolourous when mature, bluish green and glabrous on upper face, paler with a dense yellowish brown tomentum on the lower face; inflorescence composed of short racemes aggregated into axillary fascicles or ramiflorous pseudopanicles; flowers whitish or creamish; fruit stipitate, indehiscent, orbicular to subreniform with a velvety chocolate-brown tomentum.

Terra firme forest at edge of igarapé.

The fruits float and some, at least, must be water dispersed.

North side of island at Três Igarapés, ca. 1 km up igarapé from subcamp 1, *Lewis* 1556 (fr.)!

31. **CENTROLOBIUM** Mart. ex Benth.

Rudd in Journ. Wash. Acad. Sci. 44 (9): 284–288 (1954) (generic review). Cavalcante de Lima has in press a study of the Brazilian species outside the Amazon.
7 spp. in S. America.

Centrolobium paraense Tul. in Arch. Mus. Par. 4: 87 (1844). Fig. 5G.
"Pau rainha".

Tree, 10–30 m, DBH 25–80 cm; young stems, leaf and inflorescence rachises with a dense felty tomentum of greyish yellow hairs; stipules foliaceous, heart-shaped, densely sericeous on their outer faces, the hairs intermixed with numerous small orangish glands; leaves pinnate; leaflets ovate to ± oblong, the lower face moderately to densely covered with tiny flat (sometimes almost sunken) orangish glands, these easily observed with a ×10 lens; inflorescence an axillary or terminal panicle; bracts and bracteoles ± persistent, densely felty tomentose with orangish glands intermixed; calyces with a dense chocolate-brown tomentum; flowers ca. 10–15 mm long; petals yellowish tangerine, the standard with a central reddish blotch on the inner face; fruit a robust, stipitate, indehiscent, 1-seeded samara, the seed chamber basal, spiny, the spines pubescent, the wing large and papery (regularly over 10 cm long), a ± erect stylar 'spine' arising from the wing just above the seed chamber.

Terra firme forest.

A characteristic tree of the forest around base-camp, with its crown in the upper canopy or occasionally emergent. Leaflets collected from the ground can be easily recognised by their orangish glands and the fruit is quite unlike any other legume. Trees cut in early 1987, to clear various trackways, had produced lush 2 metre shoots by October the same year. On these new shoots, scale insects in the axils of stipules were being farmed by ants which responded aggressively when the stems were touched.

Estrada from base-camp to Santa Rosa, *Lewis* 1420 (fl. & fr.)!; same locality, *Lewis* 1697 (sterile)!; forest trails close to estação, *Hopkins et al.* 521 (fl. & fr.)!; same general locality, *Ratter et al.* 5489 (fr.)!; without detailed locality, *Lima* 790 (fl.)!

32. **FISSICALYX** Benth.

1 species in Venezuela, Guyana, Panama and just entering Brazil.

Fissicalyx fendleri Benth. in Journ. Linn. Soc. 5: 79 (1860). Fig. 5J.

Tree to 35 m, DBH 50 cm; bark light brown or greyish, flaky; leaves pinnate; leaflets opposite, almost entirely glabrous, acuminate; inflorescence a terminal panicle, flowering when tree almost leafless; buds golden-brown appressed-pubescent; calyx splitting spathaceously at anthesis; flowers yellow or orangish; fruit flat, indehiscent, ovate-elliptic with broad lateral wings, these with a close ± parallel venation at right angles to the margins, 1-seeded.

Terra firme forest.

Apparently rare, certainly little-known from herbarium collections.

West of Casa de Maracá, *Milliken et al.* 444 (sterile)!; Proctor plot 2 [near base-camp], tree 0082, *Ratter et al.* 5616V (sterile)!; N end of the Trilha de Preguiça, transect tree 7/3a, *Ratter et al.* 6128V (sterile)!; near base-camp, *Rodrigues* 10152 (fl.)!

33. **ABRUS** Adans.

17 spp., pantropical.

Abrus pulchellus Wall. ex Thw., Enum. Pl. Zeyl. : 91 (1859).

One subspecies occurs on the island.

subsp. **tenuiflorus** (Benth.) Verdc. in Kew Bull. 24: 250 (1970). Fig. 6E.
Abrus tenuiflorus Benth. in Mart., Fl. Bras. 15: 216 (1859).

Twiner to 2 m; leaves pinnate; leaflets opposite, dark green; inflorescence ± nodose, the flowers borne on cushion-like reduced branchlets; corolla white to purple, usually pale pink; fruit a short dehiscent pod ca. 3 cm long.

Terra firme forest in swampy areas.

Most plants observed by G.P.L. were sterile. The type, *Spruce* 786 was collected at Santarém. The closely related *A. precatorius* L., not known to occur on Maracá, contains abrin which is highly toxic if it enters the bloodstream.

Estrada from base-camp to Santa Rosa, *Lewis* 1514 (fr.)!; near base-camp, *Milliken* 799 (fl.)!; on trail to the transect of the tatu, ca. 3 km from the station, *Ratter et al.* 5382 (fr.)!

34. **LONCHOCARPUS**

Pittier in Contrib. U.S. Nat. Herb. 20: 37–93 (1917) (Middle American species).
About 100 spp. in tropical America and Africa; some workers have treated American species under the genus *Derris* (e.g. Macbride in the Flora of Peru). Brazilian species are currently being studied by de Azevedo at Campinas and the genus is also under study by Poppendieck at Hamburg and Sousa in Mexico City.

Lonchocarpus sp. aff. negrensis is not included in the key due to a lack of adequate material.

1. Trees with densely tomentose fruits:
 2. Fruit subcoriaceous, 1-seeded, the seed central, the valves acute at both ends and lacking any marginal wings ... ... **L. sp. aff. margaritensis**
 2. Fruit ± woody, (1–)several-seeded, the valves ± obtuse at both ends, irregularly undulate, the upper margin slightly winged ... ... **L. sericeus**
1. Lianas; fruits glabrous or sparsely to moderately appressed-pubescent:
 3. Fruit woody, linear-oblong, irregularly constricted between the several seeds, tending to break up along inter-seminal septa, the valves appressed-pubescent; standard densely sericeous on outer face; calyx teeth subequal ... ... **L. glabrescens**
 3. Fruit membranaceous, semi-ovate, widest near the middle, 1-seeded, the seed ± central, the valves glabrous; standard glabrous but for a few hairs near the apex on outer face; calyx teeth unequal in length, the lowest longer ... ... **L. martynii**

Lonchocarpus glabrescens Benth. in Hook. Kew Journ. Bot. 2: 233 (1850). Fig. 5A, Plate 6D and E.

Derris glabrescens (Benth.) Macbride in Fl. Peru, Bot. Series Field Mus. Nat. Hist. 13(3), no. 1: 261(1943).

"Timbó".

Robust woody climber to 5 m; bark greyish white to pale brown, lenticellate; leaves pinnate, (9–)11-foliolate; leaflets bluish green to dark green, subcoriaceous, ovate to elliptic, acute to acuminate at apex, the blade glabrous on upper face, lower face sparsely pubescent, glabrescent; inflorescence an axillary or ramiflorous, erect pseudoraceme up to 45 cm long; (4–)6(–8) flowers on each cushion-like node; calyces pinkish purple to red, appressed-pubescent, the teeth obscure; standard petal broadly orbicular, reflexed, pale lilac-pink with a central yellow blotch on the inner face, the outer face densely silky pubescent; wings and keel lilac; fruit woody, purplish brown, indehiscent but breaking up at the septa (and therefore ± a loment), sparsely to moderately appressed-pubescent, up to 24 × 2 cm, 3–11-seeded.

River margins.

Particularly in March the long, erect, pink-flowered inflorescences form a striking display along the river banks. A fish poison is extracted from the roots and used by local Indians. The fruits are quite unlike those of any other *Lonchocarpus* on the island.

Between base-camp and Maracá house, *Lewis* 1429 (fl.)!; Santa Rosa station, *Lewis* 1445 (fr.)!; down-river from base-camp towards eastern end of island, *Lewis* 1592 (fr.)!; margin of Rio Urariquera; *Lewis* 1668 (fl.)!; Furo de Maracá, *Milliken et al.* 766 (fl. & fr.)!; same locality, *Sette Silva & Lima* 5815 (fl.)!

Fig. 5., all × ⅔. **A**: *Lonchocarpus glabrescens* (from Lewis 1592); **B**: *Machaerium kegelii* (from Lewis 1564); **C**: *Machaerium sp. A* (from Lewis 1439); **D**: *Machaerium inundatum* (from Lewis 1598); **E**: *Machaerium quinata* var. *parviflorum* (from Milliken 80); **F**: *Dalbergia riedelii* (from Lewis 1556); **G**: *Centrolobium paraense* (from Lewis 1420); **H**: *Andira surinamensis*, **i**, fruit, **ii**, seed (from Milliken 307); **J**: *Fissicalyx fendleri* (from Venezuela, Fendler 2223); **K**: *Hymenolobium petraeum* (from Lewis 1698); **L**: *Vigna lasiocarpa* (from Lewis 1427); **M**: *Vigna adenantha* (from Lewis 1434); **N**: *Vigna luteola* (from Bahia, Noblick 2962) —since preparation of this fig. *V. luteola* found to have been collected off island (see text).

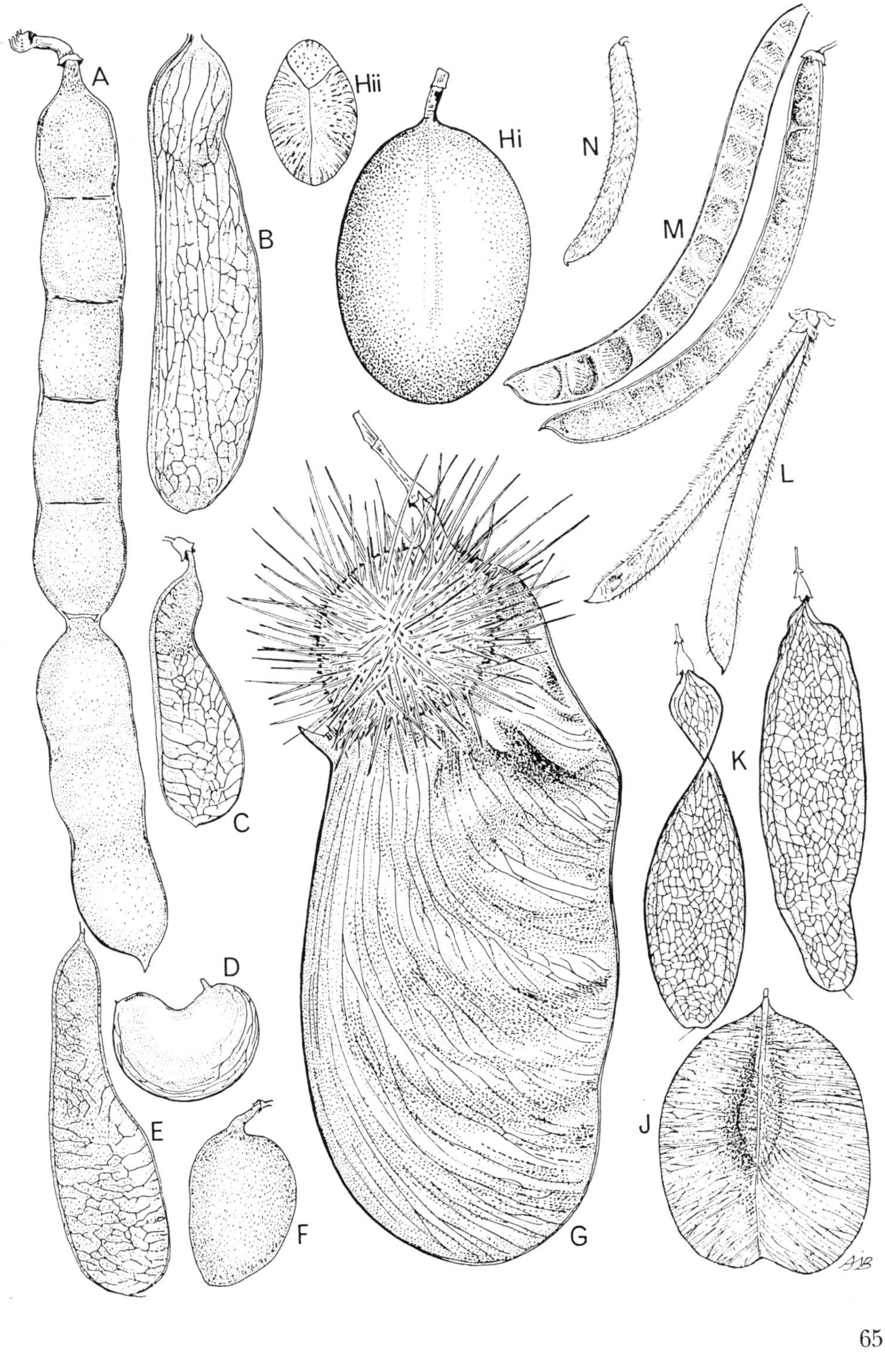
A
B
Hii
Hi
N
M
L
C
K
D
E
J
F
G

Lonchocarpus sp. aff. margaritensis Pittier in Contrib. U.S. Nat. Herb. 20: 91 (1917). Fig. 6V.

Tree 25–30 m, DBH 33–62 cm; bark pale greyish white with prominent red-brown tubercular lenticels; wood white; leaves pinnate, leaflets opposite, broadly lanceolate to ovate-elliptic, the apex mucronate, the upper face glabrous but for a few hairs on the veins, the lower face densely appressed white pubescent, the tertiary, fine-network, venation very evident; flowers unknown; fruit subcoriaceous, densely appressed-pubescent, 6–8 × 2 cm, 1-seeded, the seed ± central.

Terra firme forest; partially flooded tall forest.

Also closely related to *L. ernesti* Harms and *L. hedyosmus* Miq.

W of Casa de Maracá, *Milliken et al.* 449 (fr.)!; near Igarapé camp, Fumaça trail, transect tree no. 5/1, *Ratter et al.* 5867V (sterile)!

Lonchocarpus martynii A.C. Sm. in Amer. Journ. Bot. 24: 577 (1937).
"Cipó cururu", "Timbó".

Robust woody climber to 15 m; leaves pinnate, 7–9-foliolate; leaflets elliptic, apically acuminate, glabrous, the tertiary venation finely reticulate and evident when viewed through a×10 lens; inflorescence an erect, elongate pseudoraceme, 30–40 cm long, the deep blue, narrowly pedicellate flowers clustered onto cushion-like nodes; calyces sparsely appressed-pubescent, the teeth distinct, acute to acuminate, the lowest one slightly larger than the others; standard petal somewhat pubescent at the apex on the outer face; fruit membranaceous, glabrous, indehiscent, finely reticulate-veined, up to 9.5 × 3.5 cm, 1-seeded, the seed ± central.

Terra firme forest; shrubby riverine vegetation.

Road to Santa Rosa, *Ratter et al.* 6193 (fl.)!; Furo de Maracá near eastern tip of island, *Ratter et al.* 6299 (fr.)!

Lonchocarpus sp. aff. negrensis Benth. in Journ. Linn. Soc. 4, Suppl.: 98 (1860).
"Mututi"

Tree to 20 m, DBH 35 cm; lower part of trunk 'lobed'; bark vertically fissured; leaves pinnate; persistent filiform stipels, ca. 2 mm long, at base of leaflets; leaflets elliptic, ± acuminate, upper face glabrous, lower face very sparsely pubescent; inflorescence axillary, racemose; calyces appressed-pubescent, the distinct teeth acute; petals yellow; fruit unknown.

Terra firme forest.

More material is needed to confirm the identification of this taxon.

Santa Rosa area, ca. 4 km NE of Station buildings, *Ratter et al.* 6329 (fl.)!

Lonchocarpus sericeus (Poir.) Kunth, Nov. Gen. & Sp. 6: 383 (1824). Fig. 6W.
Robinia sericea Poir. in Lam. Encycl. 6: 226 (1804).

Tree to 20 m, DBH 30 cm; bark grey-brown, smoothish with prominent lenticels in vertical lines; leaves pinnate, 7–11-foliolate; leaflets ovate, elliptic or oblong-elliptic, obtusely acuminate at apex, glabrous on upper face, shortly pubescent on lower face; inflorescence an axillary pseudoraceme; flowers pedicellate, paired on each abbreviated node of the rachis; calyces truncate, rusty tomentose; corolla pinkish, pale purple or lilac, the standard densely silky on outer face; fruit ± woody, rusty tomentose, indehiscent, straight or ± falcate, very slightly winged along upper margin, ca. 2.5–10 × 2 cm, 1–several-seeded.

Partially flooded tall forest.

West of Casa de Maracá, *Milliken et al.* 450 (fr)!

35. **TEPHROSIA** Pers.

Over 400 spp., pantropical with most species in Africa.

Tephrosia sessiliflora (Poir.) Hassl. in Fedde Repert. 16: 162 (1919). Fig. 6D.
Cytisus sessiliflorus Poir., Encycl. Suppl. 2: 439 (1812).

Woody herb to 2 m; stems with an appressed, densely silky pubescence; leaves digitately trifoliolate, almost sessile or with a very short petiole; leaflets strongly discolourous, the upper face bluish green, the lower face silvery grey due to a dense appressed silky pubescence, narrowly oblanceolate, the apex rounded to emarginate, mucronulate; secondary venation parallel, prominent on upper face, obscured by indumentum on lower face; flowers solitary or paired in leaf axils, the pedicels shorter than petioles; calyx densely silky pubescent, the lanceolate, acuminate teeth longer than the tube; flowers pale lemon yellow, the standard densely appressed-pubescent on the outer face; fruit a dehiscent, linear pod, the sutures accentuated by an ochre-orangish brown tomentum, the valves with a dense greyish white tomentum, ca. 35–50 × 4.5 mm, ca. 5–8-seeded; seeds orangish brown with a small creamish yellow aril.

Grassy campo — savanna.

Estrada from base-camp to river, *Lewis* 1578 (fl. & fr.)!; same locality, *Milliken* 329 (fl. & fr.)!

36. **COURSETIA** DC.

Lavin in Systematic Botany Monographs 21: 167 pp. (1988) (revision of the genus).

38 spp. from southern U.S.A. through Central America to Argentina with minor range extensions in Paraguay and east-central Brazil.

Coursetia ferruginea (Kunth) Lavin in Advances in Legume Systematics 3: 63 (1987). Fig. 6B, Plate 8D.
Robinia ferruginea Kunth, Nov. Gen. & Sp. 6: 395 (1823); *Humboldtiella ferruginea* (Kunth) Harms (1923); *Coursetia arborea* Griseb. (1859); *Callistylon arboreum* (Griseb.) Pittier (1928); *Humboldtiella arborea* (Griseb.) Hermann (1948).

Shrubs to small spindly trees (forest margins and capoeira) to larger trees (in forest) (1–)2.5–10(–15) m, terminal branchlets sometimes scandent; bark smoothish, greyish white to black, lenticels numerous in longitudinal files which give rise to vertical fissures as the trunk thickens; stipules erect, acuminate, persistent, 3–4 mm long; leaves pinnate with 12–26 alternate leaflets; stipels minute; leaflets strigose to glabrescent on upper face, strigose to pilose or sericeous on lower face; inflorescences sessile racemes commonly clustered on conspicuous short shoots, sometimes pendent, flowering branches often completely leafless, inflorescence rachis and young shoots ferruginous tomentose; calyx lobes rounded to blunt, their margins white woolly, a central vein in each lobe terminating as a tiny glandular structure; flower colour various depending on habitat, in forest petals mostly white, the standard with a central basal yellow blotch on inner face, at forest margins and savanna edge the petals white tinged pink, the standard often deep rose pink to wine red on the outer face, creamy white to pinkish with a central yellow blotch on the inner face, wing petals pinkish to wine red, keel white sometimes tinged pinkish; anthers yellow; fruit a dehiscent woody pod, green tinged reddish purple, maturing brown, stipitate, the staminal sheath and upturned style persistent, ca. 10–15 cm long × ca. 7–10 mm wide, 14–18-seeded.

Terra firme forest; forest margins and surviving vegetation between estrada from base-camp to river and marshland; secondary scrub.

Flowering is mostly in the dry season (January to April) but can be sporadic through to October. The calyx-lobe glands are visited by small black ants, mainly of the genus *Crematogaster*. Several collections record inflorescences swarming with ants (e.g. *Lewis* 1442, *Ratter et al.* 5374, *Stannard & Arrais* 668). Presumably the ants protect the flowers (especially in bud) from predation by other insects, although flowers at anthesis often have their nectar taken by robber bees that drill a hole through the side of the calyx. Open flowers are visited by large black bees, mainly of the genus *Xylocopa*.

Estrada from base-camp to Rio Urariquera, *Lewis* 1425 (fl.)!; base-camp environs, *Lewis* 1442 (fl. & fr.)!; as for *Lewis* 1425, *Harley* 24728 (fl.)!; regeneration plots by base-camp, *Milliken & Miller* 772 (fl. & fr.)!; by base-camp, *Ratter et al.*5374 (fl.)!; estrada to Santa Rosa, *Ratter et al.* 5482V (sterile)!; Fumaça trail, near Igarapé camp, *Ratter et al.* 5865V (sterile)!; north end of trilha de Preguiça, near Furo de Santa Rosa, *Ratter et al.* 6124 (fl.)!; mata da picada para a lagôa [near base-camp], *Stannard & Arrais* 668 (fl. & fr.)!

37. **SESBANIA** Scop.

About 50 spp., widespread in the tropics and subtropics, generally in seasonally or permanently wet places.

Sesbania exasperata Kunth, Nov. Gen. & Sp. 6: 534 (1824). Fig. 6A.

Herb to 3 m with glabrous, pithy stems; stipules linear-lanceolate, papery, deciduous; leaves pinnate, multifoliolate; stipels tiny, under 1 mm long; leaflets linear-oblong, mucronate, discolourous, glabrous on both faces; inflorescence an axillary, few-flowered raceme; calyx campanulate, the tube glabrous, the acuminate teeth with pubescent margins; bracts and bracteoles ± as for stipules; corolla up to 3 cm long, bright yellow, the standard sometimes speckled purplish black on outer face; fruit a long, linear, ± torulose, septate, glabrous, dehiscent, many-seeded pod, the sutures somewhat thickened, the seed cavities ± oblong.

Seasonally flooded savanna.

Ilha da Nova Olinda, *Milliken & Bowles* 359 (fl. & young fr.)!

38. **INDIGOFERA** L.

700 spp. in the tropics and subtropics, principally in Africa.

Indigofera lespedezioides Kunth, Nov. Gen. & Sp. 6: 457 (1824). Fig. 6C, Plate 8E.

Perennial woody-based herb to 1 m; leaves pinnate; leaflets bluish grey-green with an appressed pubescence of characteristic medifixed (or two-branched) hairs, oblanceolate, their apices acute to obtuse to emarginate, mucronate; inflorescence an axillary raceme, usually longer than the leaf that subtends it, particularly when all flowers have reached anthesis; petals brick-red or pinkish, the staminal sheath persistent after the petals have fallen; fruit a deflexed, dehiscent, cylindrical pod, ca. 25–30 × 2 mm, ca. 8–10-seeded.

Trackside scrub, secondary growth and campo.

Estrada from base-camp to Rio Urariquera, *Lewis* 1566 (fl. & fr.)!; near base-camp, *Harley* 24700 (fl. & fr.)!; same locality, *Hopkins et al.* 869 (fl. & fr.)!; track to Rio Urariquera, *Lima & Nelson* 764 (fl. & fr.)!; near base-camp, *Ratter et al.* 5391 (fl. & fr.)!

39. **DESMODIUM** Desv.

de Azevedo, Dissertação para Mestre em Biologia, Campinas — SP: 1–315 (1981).

About 300 spp. in the tropics and subtropics with centres of great diversity in E. Asia, Mexico and Brazil.

1. Inflorescence a short, dense, terminal raceme, ca. 1–2.5 cm long; calyx long silky pilose and ciliate; primary bracts persistent ... ... **D. barbatum**
1. Inflorescence a lax raceme over 3 cm long (usually much longer) or a sparingly to many-branched panicle:
  2. Inflorescences of each main stem simple, composed of 1–2 terminal racemes:
    3. Leaflets small, orbicular, obcordate or elliptic, ca. 1.5–3 × 1–2 cm; plant prostrate with erect inflorescences ... ... **D. adscendens**
    3. Leaflets larger, generally ovate; plant erect or procumbent with some erect stems:
      4. Plants ± procumbent, rooting at the nodes, some stems erect; stipules free from each other; petioles often equalling or longer than the lateral leaflets ... ... **D. affine**
      4. Plants suffruticose, erect; stipules fused basally or overlapping; petioles usually much shorter than leaflets ... ... **D. incanum**
  2. Inflorescences of each main stem with a larger number of terminal or axillary racemes or inflorescence a simple panicle:
    5. Stipe of fruit (3–)4–8 mm long, articles ca. 4–6 mm wide, subelliptic to reniform, the upper suture not thickened; inflorescences axillary racemes, these long, erect, flowering only above the middle, the peduncle with persistent bracts a few cms from its base, these ca. 3 mm wide; leaflets usually moderately pubescent on lower face ... ... **D. axillare**
    5. Stipe of fruit 1–2 mm long, articles up to 3 mm wide, lunate, the upper suture thickened; inflorescences axillary or terminal racemes, or sparingly-branched panicles flowering along much of their length, the peduncle lacking bracts near its base or if present these inconspicuous, under 1 mm wide; leaflets very sparsely pubescent on lower face ... **D. wydlerianum**

Desmodium adscendens (Sw.) DC., Prodr. 2: 332 (1825). Fig. 6H.
Hedysarum adscendens Sw., Prodr. Veg. Ind. Occ.: 106 (1788); *Meibomia adscendens* (Sw.) Kuntze (1891).

Semi-prostrate, woody-based herb rooting at the nodes, sometimes becoming a scrambling twiner; stems with appressed white or silvery indumentum, glabrescent with age; stipules persistent, papery, striate, acuminate; leaves pinnately trifoliolate; stipels minute, filiform; leaflets broadly ovate to suborbicular, upper face sparsely pubescent to glabrous, lower face moderately appressed-pubescent; inflorescence a terminal raceme; flowers purplish pink; fruit a loment of 2–4 articles, the upper margin ± straight, the lower margin sinuate, the articles covered in uncinate hairs.

Seasonally flooded brejo forest with many palms; secondary scrub.

Estrada from base-camp to Santa Rosa, *Lewis* 1452 (fr.)!; near base-camp, *Hopkins et al.* 785 (fl. & fr.)!

Desmodium affine Schlecht. in Linnaea 12: 312 (1838).
Desmodium albiflorum Salzm. (1859); *Meibomia albiflora* (Salzm.) Kuntze (1892).

Erect herb; older stems longitudinally ribbed, glabrescent; acuminate; leaves pinnately trifoliolate; stipels filiform; leaflets broadly ovate to elliptic, upper face sparsely pubescent, lower face moderately pubescent especially on the veins; inflorescence a terminal raceme to 20 cm long; flowers pinkish mauve; fruit a loment of (5–)7(–9) articles, the upper margin ± straight, the lower margin sinuate, faces of the articles densely pubescent with uncinate hairs.

Grassy scrubland on river bank.

Probably inadvertently introduced onto the island by man in the last 10 years.

Estrada from base-camp to Rio Urariquera, *Lewis* 1575 (fl. & fr.)!

Desmodium axillare (Sw.) DC., Prodr. 2: 333 (1825). Fig. 6K.
Hedysarum axillare Sw., Prodr.: 107 (1788); *Desmodium reptans* DC. (1825); *Meibomia axillaris* (Sw.) Kuntze (1891).

Creeping herb with reddish stems, rooting at the nodes; stipules persistent, papery, striate, acuminate; leaves pinnately trifoliolate; petiole 7–8 cm long; stipels filiform, ca. 5 mm long; leaflets ovate, acuminate, the upper face glabrous, the lower face densely appressed-pubescent; inflorescence a long axillary raceme; flowers pink; fruit a loment of ca. 2 articles, these with an indumentum of uncinate hairs.

Terra firme forest on elevated ground.

Near Casa de Maracá, *Milliken* et al. 123 (fl.)!

Fig. 6. **A**: *Sesbania exasperata*, **i**, fruit × ⅔, **ii**, seed in seed chamber × 4 (both from Paraguay, Balansa 1379); **B**: *Coursetia ferruginea* × ⅔ (from Lewis 1442); **C**: *Indigofera lespedezioides*, **i**, fruit × 2, **ii**, section of lower face of leaflet showing medifixed hairs × 25 (both from Hopkins et al. 869); **D**: *Tephrosia sessiliflora* × ⅔ (from Lewis 1578); **E**: *Abrus pulchellus* subsp. *tenuiflorus* × 2/3 (from Mato Grosso, Anderson 9905); **F**: *Crotalaria micans*, **i**, fruit × ⅔, **ii**, seed × 6 (from Lewis 1572); **G**: *Soemmeringia semperflorens*, **i**, flower showing papery standard × 2, **ii**, bisected flower showing developing fruit × 2, **iii**, single article of fruit × 6 (all from Lewis 1600); **H**: *Desmodium adscendens* × 2 (from Lewis 1452); **J**: *Desmodium wydlerianum* × 2 (from Surinam, Lanjouw & Lindeman 1804); **K**: *Desmodium axillare* × 2 (from Bahia, Lewis & de Carvalho 703); **L**: *Desmodium barbatum* × 2 (from Lewis 1479); **M**: *Desmodium incanum* × 2 (from Lewis 1573); **N**: *Aeschynomene histrix* var. *histrix* × 2 (from Lewis 1569); **O**: *Aeschynomene paniculata* × 2 (from Mato Grosso, Irwin et al. 16114); **P**: *Aeschynomene sensitiva* var. *sensitiva* × 2 (from Lewis 1475); **Q**: *Aeschynomene fluminensis* var. *fluminensis* × 2 (from Hopkins et al. 887); **R**: *Stylosanthes guianensis* var. *gracilis* × 2 (from Surinam, Lanjouw & Lindeman 1604) — since preparation of this fig. *S. guianensis* found to have been collected off island; **S**: *Stylosanthes humilis* × 2 (from Lewis 1694); **T**: *Zornia sp. aff. latifolia*, **i**, fruit partially enclosed in bract × 2, **ii**, pubescent bristle from article surface × 20 (from Lewis 1599); **U**: *Macroptilium lathyroides* × ⅔ (from Guyana, Davis 298); **V**: *Lonchocarpus sp. aff. margaritensis* × ⅔ (from Milliken 449); **W**: *Lonchocarpus sericeus* × ⅔ (from Milliken 450).

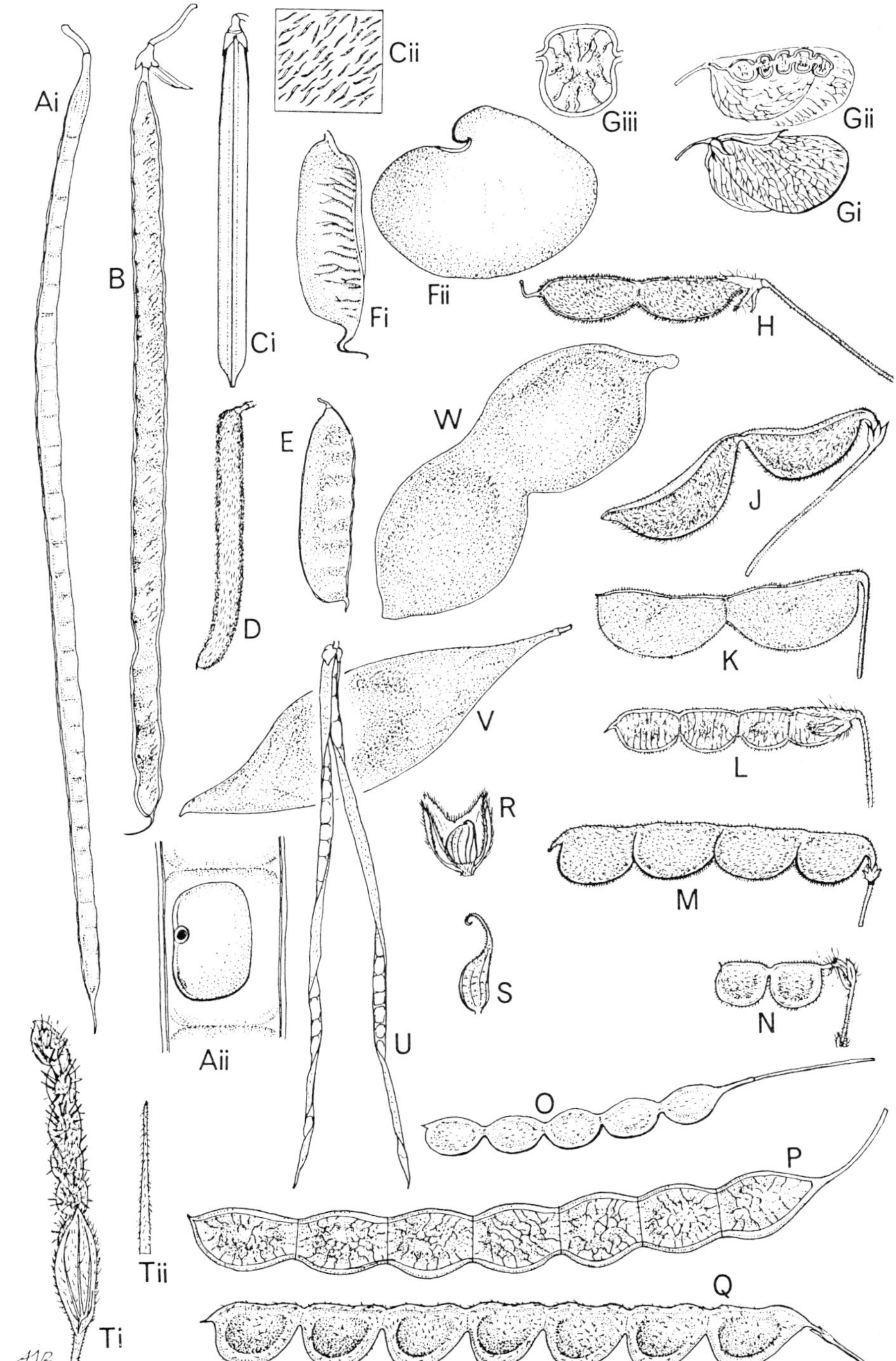
Ai
B
Ci
Cii
Fi
Fii
Giii
Gii
Gi
H
W
E
D
J
K
V
L
R
M
Aii
S
U
N
O
P
Tii
Q
Ti

Desmodium barbatum (L.) Benth. in Mart., Fl. Bras. 15(1): 95 (1859). Fig. 6L.
Hedysarum barbatum L., Sp. Pl. ed. 2: 1055 (1763); *Desmodium coeruleo-violaceum* DC. (1825); *Meibomia barbata* (L.) Kuntze (1891).

Semi-prostrate, woody-based perennial herb to 75 cm; younger stems with a dense, spreading or ± appressed silvery or yellowish pubescence; stipules persistent, papery, acuminate; leaves pinnately trifoliolate; stipels filiform; leaflets bluish green, ± oblong, upper face glabrous or glabrescent, lower face appressed-pubescent; inflorescence an axillary or terminal, densely congested raceme; the calyces with long spreading golden hairs; flowers pinkish mauve to purplish; fruit a loment of 3–6 articles, the upper margin ± straight, the lower margin ± undulate.

Campo; wet grassy savanna; scrubby transition between low campina forest and seasonally inundated campo.

A common, weedy species throughout Brazil showing much morphological variation.

Estrada from base-camp to Rio Urariquera, *Lewis* 1479 (fl. & fr.)!; near base-camp, *Milliken* 355 (fl. & fr.)!; Santa Rosa area, *Ratter et al.* 6190 (fl.)!

Desmodium incanum (Sw.) DC., Prodr. 2: 332 (1825). Fig. 6M.
Hedysarum incanum Swartz, Prodr. Veg. Ind. Occ.: 107 (1788); *Meibomia incana* (Sw.) Vail (1892); *Desmodium canum* (Gmel.) Schinz. & Thell. (1913); *Desmodium frutescens* Schindl. (1926).

Erect herb from creeping rhizome; stems sparsely pubescent, glabrescent; stipules ± persistent, papery, striate, ovate-lanceolate, acute; leaves pinnately trifoliolate; stipels filiform, ca. 1 mm long; leaflets mid-green, glabrescent on upper face, sparsely to moderately pubescent on lower face, the blade ovate to elliptic or lanceolate, sometimes ± orbicular, secondary and tertiary venation prominent on lower face, sparsely pubescent; inflorescence a terminal raceme; two distinct colour forms, flowers either cream to whitish or pinkish lilac; fruit a loment of (2–)4(–5) articles, the upper margin ± straight, the lower margin undulate, moderately pubescent with uncinate hairs.

Disturbed river banks and grassy scrubland.

Possibly introduced by man.

Estrada from base-camp to Rio Urariquera, *Lewis* 1573 (cream-flowered form) (fl. & fr.)!; same locality, *Lewis* 1574 (pink-flowered form); same general area, *Nelson & Mera* (pink-flowered form) (fl.)!; by base-camp, *Stannard & Arrais* 827 (cream-flowered form) (fl.)!

Desmodium wydlerianum Urb., Symb. Ant. 2: 302 (1900). Fig. 6J.
Meibomia wydleriana (Urb.) Britton in Sci. Surv. Porto Rico & Virgin Isl. 5: 403 (1924).

Prostrate to erect herb, rooting at the nodes; stipules ± persistent, papery, acuminate; leaves pinnately trifoliolate; stipels tiny, filiform; leaflets ovate to broadly triangular, acute, discolourous, mid-green on upper face, paler on lower face with a very sparse appressed pubescence, glabrescent; inflorescence a few-flowered, axillary raceme; flowers mauve; fruit a stipitate loment of ca. 2 articles, these uncinate-pubescent.

Terra firme forest.

Also known from Venezuela, Peru and the West Indies.

Forest trails close to base-camp, *Hopkins et al.* 627 (NY, photo K) (fl.)!

40. **MUCUNA** Adans.

Burkart in Darwiniana 16: 175–181 (1970). The genus has been revised for many regions of the Old World in recent years and Windler is now studying the neotropical species. Verdcourt and Wilmot-Dear, at Kew, are currently preparing a monograph of the genus worldwide.
About 100 spp., tropics and subtropics.

Mucuna urens (L.) Medik. in Vorles. Churpf. Ges. 2: 399 (1787). Fig. 7A.
Dolichos urens L., Syst. Nat. ed. 10, 2: 1162 (1759).

Robust woody climber drying black; leaves pinnately-trifoliolate; leaflets ± ovate, acuminate, glabrous; inflorescence an axillary, pendulous pseudoraceme up to 150 cm long, the peduncle less robust than the zig-zag rachis, the flowers 2–3 per cushion-like node; calyx appressed-pubescent with some golden-yellow, irritant hairs intermixed; corolla purplish; fruit a dehiscent pod, ca. 10–20 × 2–2.5 cm, transversely lamellate, pilose with brown hairs and golden irritant hairs intermixed, one margin with 2 longitudinal, undulate wings, each about 3 mm wide, 1–4-seeded; seeds depressed-globose, nearly circled by hilum.

Riverside vegetation.

Furo de Maracá, *Milliken et al.* 436 (fl.)!

41. **DIOCLEA** Kunth

Burkart in Darwiniana 16: 181–194 (1970); Maxwell, PhD dissertation, Southern Illinois University: xvi, 1–431 (1969) (New World species; includes keys), and in Ann. Miss. Bot. Gard. 67 (3): 662–675 (1980).
About 40 spp., mostly neotropical.

1. Pod woody, apically obtuse, ca. 10–15 × 5–6 cm, glabrescent, the upper margin strongly thickened, ca. 1 cm across, 2–4-seeded; seeds very hard, subglobose, ⅔ surrounded by a linear hilum; leaflets glabrous but for a few hairs on the veins of lower face ... ... **D. violacea**
1. Pod subligneous, apically acute, ca. 9–10.5 × 1.5 cm, tomentose, the upper margin ca. 4 mm across, 7–10-seeded; seeds laterally compressed, less than ½ surrounded by hilum; leaflets sparsely to densely pubescent on lower face, sparsely pubescent on upper face:
   2. Calyx pubescent outside; bracteoles ovate to broadly lanceolate with acute to acuminate apices, ca. 1–3 × 1–2 mm ... ... **D. guianensis**
   2. Calyx glabrous outside; bracteoles suborbicular with rounded to broadly obtuse apices, ca. 6–10 × 4–6 mm ... ... **D. virgata**

Dioclea guianensis Benth., Comm. Leg. Gen.: 70 (1837). Fig. 7K.

Robust woody climber or less frequently a herbaceous twiner, to 1.5–8 m; leaves pinnately trifoliolate; stipels tiny, filiform; leaflets greenish yellow, softly pubescent, especially on lower face, abruptly acute to acuminate apically, mucronate; inflorescence an erect, axillary pseudoraceme, long-pedunculate, the flowers clustered onto persistent cushion-like nodes; buds acuminate; sepals mauve; bracteoles ovate, tardily deciduous; flowers pale mauve to bluish purple, the standard with a basal, central yellow blotch on the inner face; filaments white with purple tips; anthers yellow; fruit a laterally compressed, subligneous, dehiscent pod, the valves softly golden-tomentose, the upper margin thickened, ± 'winged', ca. 9–10.5 cm long, 7–10-seeded; seeds mid-brown with darker mottling, ca. 8 × 5 mm.

Campo; savanna; open areas in secondary forest; campo—forest boundary; swampy areas.

A common species found in several habitats. *Milliken & Bowles* 273 collected up-river from the island at Igarapé Bubu is a more robust form of *D. guianensis*, perhaps referable to *D. macrantha* Huber, and might be expected at the western end of the island.

Estrada from Santa Rosa to base-camp, *Lewis* 1446 (fl. & fr.)!; estrada from base-camp to Rio Urariquera, *Lewis* 1482 (fl.)!; north side of island, igarapé opposite Rio Trairão, *Lewis* 1546 (fl. & fr.)!; trail behind Casa Santa Rosa, *Hopkins et al.* 600 (fl.)!; without detailed locality, *Lima & Nelson* 773 (fl.)!; near base-camp, *Milliken & Lima* 43 (fl. & fr.)!; by Rio Urariquera, *Ratter et al.* 5838 (fl.)!; Fumaça trail near Igarapé camp, *Ratter et al.* 5909 (fl.)!; Santa Rosa area, ca. 4 km NE of Station buildings, *Ratter et al.* 6157A; near base-camp, *Stannard & Arrais* 803 (fl. & fr.)!

Dioclea violacea Mart. ex Benth., Comm. Leg. Gen.: 69 (1837). Fig. 7B.
Dioclea paraguariensis Hassl. (1919).

Woody climber to 10 m; exudate from cut stems drying brick-red; stipules with a large basal appendage, pilose; leaves pinnately trifoliolate, stipels filiform; pulvinules golden spreading pubescent; leaflets bottle-green, coriaceous, glabrous on both faces but for a few hairs on the veins; inflorescence a robust, erect, axillary, pseudoraceme, dark brown to golden tomentose, the flowers clustered on short-stalked cushion-like nodes; buds blackish purple; calyx with lowest lobe longer than the others; bracteoles orbicular, at length deciduous; standard deep mauvish purple with a central yellow blotch on inner face; wings purple with white flash; keel creamish, pinkish mauve; developing ovary densely golden tomentose; fruit drying blackish, a ± dehiscent, woody pod, ca. 10–15 × 5–6 cm, glabrescent, the upper margin strongly thickened, ca. 1 cm across with a lateral woody ridge either side of the suture, 2–4-seeded; seeds pebble-like, very hard, pale brown mottled black, ⅔ surrounded by a linear hilum.

Riverine forest.

The inflorescences of the specimen cited below were covered in red ants visiting the pedicel bases where there seem to be rings of nectarial tissue on the woody nodes. This is one of the species of *Dioclea* and *Mucuna* with "drift-seeds" able to float due to buoyancy chambers formed as the seed dries. They remain viable after long periods in sea water.

Between base-camp and Maracá house, *Lewis* 1436 (fl. & fr.)!

Dioclea virgata (L.C. Rich.) Amshoff in Meded. Bot. Mus. Herb. Rijks. Univ. Utrecht 52: 59 (1939).
Dolichos virgatus L.C. Rich. in Act. Soc. Hist. Nat. Paris 1: 111 (1792); *Dioclea lasiocarpa* Mart. ex Benth. (1837).

Woody climber; leaves pinnately trifoliolate; stipels minute; leaflets membranaceous to subcoriaceous, sparsely to moderately appressed-pubescent on both faces, glabrescent on upper face; inflorescence an erect axillary or terminal, robust or slender pseudoraceme, the flowers clustered on subsessile cushion-like nodes; pedicels long and slender, ca. 10–12 mm long; bracteoles ±

ovate or orbicular, papery, 5–10 mm long, completely enveloping the young buds, at length deciduous; flowers pleasantly scented, mauve, the standard with a central, basal yellow blotch surrounded deep purple on the inner face; fruit ± as for *D. guianensis*.

Riverine vegetation.

Near Cachoeira de Prumama [= Pururamé], Furo de Santa Rosa, *Milliken et al.* 720 (fl. & fr.)!

42. **CANAVALIA** DC.

Sauer in Brittonia 16: 106–181 (1964) (revision).
50 spp., pantropical with most species in the neotropics.

Upper calyx lobes rounded, very sparsely pubescent, glabrescent; standard auricled at the base above the claw; keel boomerang-shaped, obtuse, keel and wings not twisted; mature pods glabrescent, over 2 cm wide **C. dictyota**
Upper calyx lobes acute, densely white pubescent; standard without auricles; keel falcate, rostrate, keel and wings spirally twisted; mature pods pubescent, at length glabrescent; ca. 1.5 cm wide ... ... **C. grandiflora**

Canavalia dictyota Piper in Contrib. U.S. Nat. Herb. 20: 574 (1925).
Canavalia amazonica Piper (1925).

Climbing vine; leaves pinnately trifoliolate; inflorescence a pseudoraceme, the flowers clustered on persistent fleshy cushion-like nodes, calyx with the two vexillary lobes much enlarged, several times the size of the other three; flowers resupinate, petals pink, the standard with a basal yellow-green blotch on the inner face; fruit a ± woody dehiscent pod with a prominent lateral ridge on each valve, 3–5 mm below the upper margin.

Scrubby vegetation in seasonally flooded vazante.

Just inland of the Furo de Santa Rosa at approx. 3°28'N, 61°30'W, *Ratter et al.* 6199 (fl.)!

Canavalia grandiflora Benth. in Ann. Wien. Mus. Nat. 2: 135 (1837) Fig. 7L, Plate 8A.
Canavalia albiflora Ducke (1922); *Wenderothia grandiflora* (Benth.) Piper (1925).

Climbing vine to 6 m; young stems longitudinally ribbed and sparsely clothed with golden spreading hairs; leaves pinnately trifoliolate; petioles up to 11.5 cm long; leaflets ovate, sparsely appressed-pubescent on lower face, tapering apically to an acuminate drip-tip, basal pair of leaflets asymmetric about the midvein; inflorescence an axillary pseudoraceme to ca. 30 cm long; flowers in clusters on cushion-like nodes, delicately scented, somewhat fleshy; calyx with the two vexillary lobes larger than the other three, all white sericeous; petals white, the standard with a central yellowish green blotch surrounded by a mauve flash; fruit a dehiscent pod with a prominent lateral ridge on each valve ± 5 mm below the upper margin, the valves glabrescent at maturity.

Riverine forest.

Banks of the Rio Urariquera at commencement of estrada to base-camp, *Lewis* 1669 (fl.& fr.)!

43. **CALOPOGONIUM** Desv.

8 spp. in the neotropics; both Maracá species have blue flowers, a rare flower colour for Maracá legumes.

Calyx, stems and fruits covered with rust-coloured spreading hairs 2–3 mm long; fruits 3.5–5 mm wide ... ... **C. mucunoides**
Calyx, stems and fruits with appressed or spreading hairs up to 1 mm long; fruits 6–9 mm wide ... ... **C. caeruleum**

Calopogonium caeruleum (Benth.) Sauv. in Ann. Acad. Havana 5: 337 (1869). Fig. 7F.
Stenolobium caeruleum Benth., Comm. Leg. Gen.: 61 (1837).

Twining climber to 2 m; stems with spreading golden yellow or whitish hairs; leaves pinnately trifoliolate; leaflets broadly ovate, the basal pair asymmetric about the midvein, pubescent on both faces but more densely so below with the indumentum evident to the naked eye and appearing as though brushed ± parallel to the secondary veins; inflorescence an axillary or terminal pseudoraceme with the flowers clustered on small fleshy cushion-like nodes, these nodes persisting after the flowers have fallen; flowers pale blue, the standard with a central yellow blotch surrounded by lilac; fruit a pubescent, 5–7-seeded dehiscent pod with the valves indented between the seeds at right angles to the sutures.

Riverine forest.

Between base-camp and Maracá house, *Lewis* 1435 (fl. & fr.)!; same general locality, *Lewis* 1506 (fl. & fr.)!

Calopogonium mucunoides Desv. in Ann. Sci. Nat. 9: 423 (1826). Fig. 7E.

A thin-stemmed twining climber with most parts of the plant covered in spreading golden hairs; leaves pinnately trifoliolate, the terminal leaflet ovate, the basal pair asymmetrical about the midvein; both faces of the leaflets densely clothed in appressed whitish hairs, the blade apex with a short, curved mucro; inflorescence axillary, much shorter than the leaf that subtends it; calyx lobes acuminate, ca. 4 mm long; flowers blue; fruit a densely spreading pubescent, 6–8-seeded, dehiscent pod, constricted between the seeds; seeds orangish brown.

Campo.

A common weedy species in many parts of Brazil but only known from Maracá by one collection and this near to base-camp and perhaps, therefore, inadvertently introduced by man after 1978.

Near base-camp, *Lewis* 1467 (fr.)!

Fig. 7., all × ⅔. **A**: *Mucuna urens* (from Roraima Territory, Prance 10722); **B**; *Dioclea violacea* (from Lewis 1436); **C**: *Rhynchosia shomburgkii* (from Lewis 1656); **D**: *Eriosema simplicifolium* var. *simplicifolium* (from Lewis 1477); **E**: *Calopogonium mucunoides* (from Lewis 1467); **F**: *Calopogonium caeruleum* (from Lewis 1506); **G**: *Centrosema plumieri* (from Lewis 1576); **H**: *Centrosema brasilianum* var. *brasilianum* (from Lewis 1465); **J**: *Centrosema macrocarpum* (from Lewis 1476); **K**: *Dioclea guianensis* (from Lewis 1446); **L**: *Canavalia grandiflora* (from Lewis 1669); **M**: *Clitoria falcata*, **i**, before dehiscence (from Bahia, Pereira s.n.), **ii**, valves after dehiscence (from Lewis 1458).

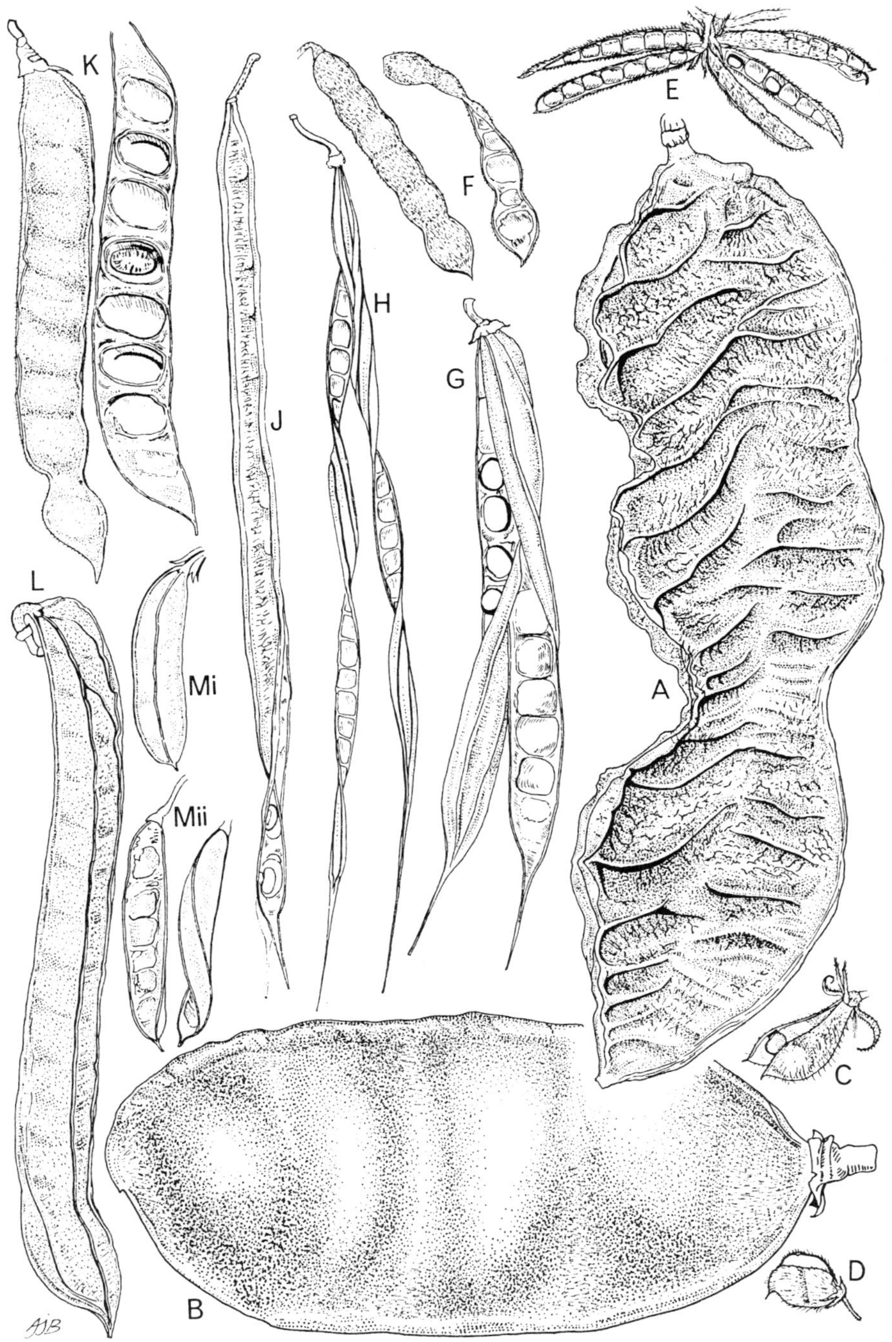
K
E
F
H
J
G
A
L
Mi
Mii
C
D
B
AJB

44. **CENTROSEMA** (DC.) Benth.

Barbosa-Fevereiro in Rodriguésia 42: 159–219 (1977) (Brazilian spp.). The genus is being revised by Williams & Clements (BRI).

45 spp., neotropics and neosubtropics.

1. Bracteoles not concealing calyx lobes, the lobes, especially the lowest one, longer than calyx tube; inner face of standard petal pale marzipan with central green blotch surrounded by purplish red band; seeds ± kidney-shaped, laterally compressed, yellowish brown without darker mottling ... ... **C. macrocarpum**
1. Bracteoles concealing calyx and lobes, especially in bud, the lobes usually shorter than the calyx tube; inner face of standard not coloured as above; seeds oblong-cylindric to subglobose, mid-brown, sparingly to moderately mottled black:
   2. Standard mainly bluish-lilac on inner face; fruit 3–4 mm wide; seeds ± oblong, ca. 4 × 2 × 1.5 mm; leaflets narrowly lanceolate to elliptic ... ... **C. brasilianum** var. **brasilianum**
   2. Standard mainly magenta on inner face; fruit 9–12 mm wide; seeds ca. 6–7 × 5 × 3 mm; leaflets broadly ovate to rhomboid ... ... **C. plumieri**

Centrosema brasilianum (L.) Benth., Comm. Leg. Gen.: 54 (1837).
Only the typical variety occurs on the island.

var. **brasilianum** Fig. 7H.
Clitoria brasilianum L., Sp. Pl.: 753 (1753); *Bradburya brasiliana* (L.) O. Kuntze (1891).

Twiner; stems with sparse pubescence of white spreading hairs, glabrescent; stipules striate, persistent; leaves pinnately trifoliolate; stipels filiform; leaflets lanceolate, glabrous; inflorescence an axillary, 1–2-flowered, reduced raceme; bracts and bracteoles foliaceous, striate; calyx with lower tooth shorter than to slightly longer than the tube, upper and lateral teeth much shorter than tube, all hidden by bracteoles in bud; flowers resupinate, bluish lilac to mauve, the standard with a central greenish-yellow blotch on inner face; fruit an explosively dehiscent pod, ca. 5 mm wide, tapering apically to a 20–40 mm long narrow beak, the valves twisting at dehiscence, these each with 2 parallel ridges running down their entire length and drying darker than the rest of the valve surface, ca. 20-seeded; seeds brown mottled greyish white.

Riverine forest; campo; scrubby transition zone between low campina forest and seasonally inundated campo.

Near base-camp, *Lewis* 1465 (fl. & fr.)!; Furo de Maracá, riverine islet in midstream, *Lewis* 1672 (fl. & fr.)!; Santa Rosa area, ca. 4 km NE of Station buildings, *Ratter et al.* 6178 (fl. & fr.)!

Centrosema macrocarpum Benth. in An. Nat. Hist. 3: 436 (1839). Fig. 7J.
Bradburya macrocarpa (Benth.) O. Kuntze (1891).

Twiner with sparsely pubescent stems; stipules and stipels striate; leaves pinnately trifoliolate; leaflets ovate, upper face glabrescent, lower face glabrous; inflorescence a few-flowered axillary raceme; lowest calyx lobe much longer than the others, exserted from bracteoles in bud; flowers resupinate; standard petal greenish yellow on outer face, inner face pale marzipan with a pale green central blotch surrounded by a purplish red band; wing petals creamish tinged pale tangerine-pink with pink blotches near the upper margin; fruit a compressed dehiscent pod, ca. 7 mm wide, tapering apically to a beak ca. 15 mm long, the valves each with 2 submarginal parallel ridges running their entire length, these connected by a fine reticulate venation and not drying significantly darker than the rest of the valves, the valves twisting at dehiscence to release ± 15–20 ochre-brown seeds.

Campo — savanna margin.

Near base-camp, *Lewis* 1476 (fl. & fr.)!

Centrosema plumieri (Turp. ex Pers.) Benth., Comm. Leg. Gen.: 54 (1837). Fig. 7G, Plate 8B.
Clitoria plumieri Turp. ex Pers., Syn. Pl. 2(2): 303 (1807); *Bradburya plumieri* (Turp. ex Pers.) O. Kuntze (1891).

Twiner to 3 m; stems glabrous; stipules and stipels striate; leaves pinnately trifoliolate; leaflets broadly ovate, glabrous but for a few hairs on the veins; inflorescence a 1-few-flowered axillary raceme; bracts and bracteoles striate, completely concealing the ± truncate calyx especially in bud; flowers resupinate, the standard greenish white on outer face, magenta-mauve on inner face darkening towards a central yellowish green vertical band; keel and wing petals white tinged magenta at apex; whole flower drying black; fruit a dehiscent pod, 10–11 mm wide, tapering apically to a beak ca. 15 mm long, each valve with two raised parallel ridges, these ca. 2.5 mm from the margins, 12–15-seeded, the seeds subglobose, orangish brown mottled black.

Trackside scrub.

A showy species only collected along the track to base-camp and perhaps inadvertently introduced by man.

Estrada from base-camp to Rio Urariquera, *Lewis* 1576 (fl. & fr.)!

45. **CLITORIA** L.

70 spp., pantropical, mostly neotropical.

Clitoria flacata Lam., Encycl. Méth. Bot. 2: 51 (1786). Fig. 7M.
Clitoria rubiginosa Juss. ex Pers. (1807); *Clitoria glycinoides* DC. (1825).

Twiner; stems sparsely to moderately pubescent; leaves pinnately trifoliolate; stipules and stipels striate, persistent; leaflets elliptic to broadly lanceolate, semi-coriaceous, discolourous, upper face dark green, lower face paler, rounded, retuse or acute at apex; inflorescence a few-flowered axillary raceme or flowers solitary in the leaf axils; peduncle equalling or often longer than the petiole and leaf rachis together; petals white with purple nectar guides on inner face of standard; fruit a stipitate, dehiscent pod, 4–5 cm long, a central rib along the face of each valve, 5–8-seeded; seeds reddish brown, subglobose, embedded in pithy tissue.

Campo; seasonally flooded savanna and damp open marsh.

By base-camp, *Lewis* 1458 (fr.)!; same locality, *Lewis* 1655 (fr.)!; same locality, *Hopkins et al.* 898 (NY, photo K) (fl. & fr.)!; same locality, *Milliken* 213 (fl. & fr.)!

46. **VIGNA** Savi

Maréchal, Mascherpa & Stainier in Boissiera 28: 160–231 (1978).
150 spp., pantropical, mostly paleotropical.

Flowers white, the standard with a central yellow blotch; fruit sparsely spreading pubescent with white hairs, glabrescent; dehiscing to reveal a silvery white non-pithy endocarp strongly contrasting with the brown exocarp ... ... **V. adenantha**
Flowers yellow; fruit densely spreading pubescent with golden yellowish brown or rusty brown hairs (these clearly visible to the naked eye), dehiscing to reveal a white pithy endocarp ... ... **V. lasiocarpa**

Vigna adenantha (G.F. Meyer) Maréchal, Mascherpa & Stainier in Taxon 27: 202 (1978). Fig. 5M, Plate 7D & F.
Phaseolus adenanthus G.F. Meyer in Prim. Fl. Esseq.: 239 (1818); *Phaseolus truxillensis* Kunth (1823); *Phaseolus latifolius* Benth. (1837); *Phaseolus radicans* Benth. (1837).

Climber with hollow stems and an indumentum of spreading hairs; stipules ± lanceolate-ovate, not appendaged below point of attachment, at length deciduous; leaves pinnately trifoliolate; stipels present; leaflets obliquely ovate to narrowly trapesiform, sparsely appressed-pubescent on both faces, the basal pair asymmetric about the midvein; inflorescence an axillary, long-pedunculate pseudoraceme, the flowers clustered on persistent, cushion-like nodes; petals white, the standard with a central yellow blotch on the inner face, the wings with a mauve flash; keel twisted; fruit a dehiscent, chartaceous, slightly falcate pod, pubescent with spreading hairs, glabrescent, outer face of valves drying brown in contrast to the shiny, silvery-white inner face, ± 14–16-seeded.

Riverine vegetation.

Between base-camp and Maracá house, *Lewis* 1434 (fl. & fr.)!

Vigna lasiocarpa (Mart. ex Benth.) Verdc. in Kew Bull. 24: 539 (1970). Fig. 5L, Plate 7B & C.
Phaseolus lasiocarpus Benth., Comm. Leg. Gen.: 76 (1837); *Phaseolus hirsutus* Benth. (1837).

"Feijão bravo".

Scrambling climber to 2 m with hollow stems and a dense indumentum of golden-yellow spreading hairs; stipules with a short appendage below the point of attachment but this often obscured by the dense pubescence; leaves pinnately trifoliolate; stipels present; leaflets obliquely ovate to rhombic or trapesiform, the margins often somewhat undulate or irregularly lobed, both faces densely velvety tomentose or moderately appressed-pubescent, the basal pair asymmetric about the midvein; inflorescence an axillary, long-pedunculate pseudoraceme, the flowers clustered on flattish nodes near the inflorescence apex; petals pale to bright yellow, the keel a spiral tube enclosing the stamens and pistil; fruit a densely spreading-pubescent dehiscent pod, the hairs golden-yellowish brown or rusty brown, the valves twisting at dehiscence to expose a somewhat pithy endocarp, ± 8–10(–12)-seeded, seeds olive-brown marbled black.

Secondary scrub; flooded campo; várzea; vazante.

da Silva et al. 2384 report many bees of genus *Trigona* visiting the 'extra floral nectaries between the fruits', presumably glandular tissue at the bases of the pedicels although this is not evident on dried material.

Estrada from base-camp to river, *Lewis* 1427 (fl. & fr.)!; near base-camp, *Hopkins et al.* 784 (fl. & fr.)!; N of Casa de Maracá, *Milliken et al.* 148 (fl.)!; trilha de Preguiça, 2–3 km N of the Casa de Maracá, *Ratter et al.* 6229 (fl. & fr.)!; estrada from base-camp to river, *da Silva et al.* 2384 (fl. & fr.)!

Vigna luteola (Jacq.) Benth. is to be expected at the western end of the island; *Milliken & Bowles* 266 is a collection of this species from further up-river. (See Fig. 5N).

47. **MACROPTILIUM** (Benth.) Urban

Maréchal, Mascherpa & Stainier in Boissiera 28: 151–157 (1978); Barbosa-Fevereiro, Master's Dissertation, Rio de Janeiro: 1–160 (1979) (Brazilian species, including maps and key).
About 20 spp. in the neotropics and neosubtropics, 10 spp. in Brazil.

Macroptilium lathyroides (L.) Urban, Symb. Antill. 9(4): 457 (1928). Fig. 6U.
Phaseolus lathyroides L., Sp. Pl. ed. 2, 2: 1018 (1763); *Phaseolus semierectus* L. (1767); *Macroptilium lathyroides* (L.) Urban var. *semierectum* (L.) Urban (1928).

Herb to 80 cm with hollow stems clothed in whitish yellow patent hairs; stipules lanceolate, acuminate, striate, up to 1 cm long, often persisting after the leaves have fallen; leaves pinnately trifoliolate; stipels filiform, ca. 2 mm long; leaflets lanceolate, acute, shortly mucronate, glabrous on upper face, appressed-pubescent on lower face; inflorescence an axillary raceme, several times longer than the leaf that subtends it, the flowers aggregated towards the apex of a long peduncle; bracteoles acuminate; calyx appressed-pubescent, the teeth acuminate; corolla purple, the wing and keel petal blades tinged greenish and the claws whitish; fruit a linear, dehiscent pod, appressed-pubescent, ca. 18–30-seeded.

Secondary scrub.

Probably introduced onto the island since development of the field station in 1978.

Along track between base-camp and river, *Milliken* 338 (fl.)!

48. **RHYNCHOSIA** Lour.

Grear in Mem. N.Y. Bot. Gard. 31 (1): 1–168 (1978) (New World revision, including keys).
200 spp., pantropical.

Rhynchosia schomburgkii Benth. in Mart., Fl. Bras. 15(1): 203 (1859). Fig. 7C.

Woody herb or shrublet, 1.5–2.5 m, sometimes with scandent branchlets, occasionally a sprawling climber; stems, petioles, inflorescence rachises and calyces densely pubescent with short white hairs and golden-yellow glandular hairs intermixed; stipules lanceolate, acuminate, papery, pubescent, striate, drying a brownish colour, 1–1.2 cm long, at length deciduous; leaves pinnately trifoliolate; stipels linear, falcate, pubescent, 4–6 mm long, persistent; leaflets yellowish green, membranaceous, narrowly ovate to ovate, the basal pair asymmetric about the midvein, the upper face finely appressed-pubescent (the hairs just visible with a × 10 lens) the pubescence denser and spreading on the main veins and margin, the lower face densely pubescent, especially on the veins, orangish red gland dots scattered amongst the indumentum and sometimes obscured by it; inflorescence an axillary or terminal raceme; calyx lobes much longer than tube, the lowest one falcate, longer than the others; petals yellow; fruit a sharply-beaked, 1–2(–3)-seeded, dehiscent pod, the valves with a moderately dense white pubescence with longer, golden-yellow, glandular-based hairs intermixed; seeds brown mottled yellow.

Transition between savanna and forest; damp *Curatella americana* savanna.

This species is characteristic of a 5–8 metre band of vegetation transitional between the forest and savanna near base-camp. Individual plants growing 2 to 5 metres apart are scattered amongst low tussock grasses just at the edge of the forest. Their root system is shallow, to about 30 cm, radiates horizontally and produces root nodules. The transitional vegetation occurs where seasonal flooding still exerts some environmental control; as the slope increases by 15–20 degrees there is an abrupt change to forest and *R. schomburgkii* is no longer to be found. There might be a connection between the activity of the rhizobial bacteria in the root nodules of this plant, the seasonal flooding of the savanna and the presence of the species in such a narrow vegetation zone. Further investigation is necessary.

Near base-camp, *Lewis* 1487 (fl.)!; same locality, *Lewis* 1656 (fr.)!; Santa Rosa, *Milliken* 365 (fl.)!

49. **ERIOSEMA** (DC.) G.Don

Grear in Mem. New York Bot. Gard. 20(3): 1–98 (1970) (revision of the American species).

130 spp., pantropical.

Leaves unifoliolate; inflorescence 2–10-flowered; flowers 12–15 mm long ... ... **E. simplicifolium** var. **simplicifolium**
Leaves trifoliolate; inflorescence ca. 20-flowered; flowers 5–7.5 mm long ... ... **E. violaceum**

Eriosema simplicifolium (Kunth) G. Don, Gen. Hist. 2: 348 (1832).
Only the typical variety occurs on the island.

var. **simplicifolium** Fig. 7D, Plate 7E.
Glycine simplicifolia Kunth, Nov. Gen. & Sp. 6: 419 (1823).

Prostrate to semi-scandent herb arising from a long, woody tap-root; stems with spreading yellowish or whitish hairs; stipules persistent, papery, striate, acuminate; leaves unifoliolate, subsessile, the petiole and pulvinus together ca. 2 mm long, lanceolate to ± ovate, the upper face glabrescent, the lower face yellow or rufous pilose, at length glabrescent, obscurely yellow gland-dotted (glands just visible with a × 10 lens); inflorescence 2–10-flowered, fasciculate or the flowers clustered at the apex of a distinct peduncle; corolla bright yellow, 12–15 mm long; fruit a chartaceous, ellipsoid, 2-seeded, dehiscent pod, ca. 15 mm long, the valves yellow-pilose; seeds olive-brown mottled purplish black, arillate.

Campo and recently burnt savanna on greyish white sandy soil.

Estrada from base-camp to Santa Rosa, *Lewis* 1444 (fl. & fr.)!; by base-camp, *Lewis* 1477 (fl. & fr.)!; estrada to Santa Rosa, *Lima & Nelson* 739 (fl.)!; Santa Rosa, *Milliken* 387 (fl.)!

Eriosema violaceum (Aubl.) G. Don, Gen. Hist. 2: 347 (1832).
Cytisus violaceus Aubl., Hist. Pl. Guian. 2: 766 (1775).

Erect herb to subshrub to 1 m; stems appressed-pubescent with yellow or whitish hairs; stipules persistent, lanceolate, connate, papery, striate, acuminate; leaves pinnately trifoliolate, subsessile, the petiole and pulvinus together 1–2 mm long; leaflets linear-oblong, ± bullate, their margins revolute, shortly villous on both faces (the lower face with very obscure sunken gland dots); inflorescence a shortly-pedunculate axillary raceme of ca. 20 flowers; corolla yellow, 5–7.5 mm long; fruits ± as for *E. simplicifolium*; seeds black or dark greyish mottled black.

Damp campo.

Near base-camp, *Lewis* 1459 (fl. & fr.)!; same locality, *Lewis* 1489 (fl. & fr.)!; same locality, *Lewis* 1657 (fl.)!; same locality, *Ratter et al.* 5386 (fl. & fr.)!

50. **AESCHYNOMENE** L.

Rudd in Contribs. U.S. Nat. Herb. 32(1): 1–172 (1955) (revision of the American species), and in Journ. Wash. Acad. Sciences 49(2): 45–52 (1959), and in Phytologia 15(2): 114–119 (1967) (supplementary notes).
About 150 spp. in the tropics to warm temperate America, Africa & Asia.

1. Stipules not appendaged below their point of attachment; plants of sandy savanna and grassy campo:
  2. Inflorescence paniculate, usually terminal, much larger than leaves; stipules inconspicuous; fruit moniliform, 4–6-articulate, glabrous at maturity; whole plant ± glabrescent ... ... **A. paniculata**
  2. Inflorescence a congested, few-flowered axillary raceme, much shorter than the leaf that subtends it; stipules conspicuous, lanceolate, acuminate, ca. 10 mm long; fruit of 2(–3) pubescent articles, the upper margin straight, the lower one sinuate; whole plant with stiff, spreading glandular hairs ... ... **A. histrix** var. **histrix**
1. Stipules appendaged below their point of attachment; plants growing in water or seasonally inundated areas:
  3. Whole plant drying blackish; articles of fruit glabrous or very sparsely hispid; stipe of fruit 6–8 mm long ... ...**A. sensitiva** var. **sensitiva**
  3. Whole plant drying greenish or brownish with grey stems; articles of fruit with short bristles or crisped hairs; stipe of fruit 1–2 mm long ... ... **A. fluminensis** var. **fluminensis**

Aeschynomene fluminensis Vell., Fl. Flum.: 310 (1825).
Only the typical variety occurs on the island.

var. **fluminensis** Fig. 6Q.

Woody herb to 1.5–ca. 3 m; leaves pinnate; stipules persistent, appendaged at the base, the leaflets narrowly elliptic, their midveins extending beyond leaflet apex as a short mucro; inflorescences short, few-flowered axillary racemes or the flowers in pairs in leaf axils; petals yellow, the standard brighter than the other 4; fruit a loment of (3–) 5–7 one-seeded articles, these clothed in short bristles or crisped hairs.

Damp open marsh; seasonally flooded campo.

A wide ranging taxon, the type of which was collected in Rio de Janeiro state.

In front of the estação, *Hopkins et al.* 887 (fl. & fr.)!; same locality, *Ratter et al.* 5577 (fl. & fr.)!; same locality, *Ratter et al.* 5368 (fl. & fr.)!

Aeschynomene histrix Poir. in Lam. Encycl. Suppl. 4: 77 (1816).
Only the typical variety is known to occur on the island.

var. **histrix** Fig. 6N.

Woody-based herb to ± 80 cm; stems clothed with spreading hairs; stipules striate, acuminate, not appendaged below their point of insertion; leaves pinnate; leaflets yellowish-green, secondary venation prominent on the lower face; inflorescences congested, few-flowered axillary racemes with persistent bracts, several times shorter than the leaves that subtend them; flowers pale yellow; fruit a loment of 2(–3) articles, these with a fine white pubescence.

Savanna on white sand; trackside scrub.

Another widely dispersed taxon throughout Central and South America, the type of which was collected by Martin in French Guiana.

Estrada from base-camp to Rio Urariquera, *Lewis* 1569 (fl. & fr.)!; estrada from estação sede to Santa Rosa, *Lewis* 1666 (fl. & fr.)!

Aeschynomene paniculata Willd. ex Vog. in Linnaea 12: 95 (1838). Fig. 6 O.

Erect woody-based spindly herb to 2 m; stems longitudinally ribbed; stipules striate, acuminate, not appendaged at their bases; leaves pinnate; leaflets bluish green; inflorescences large, usually terminal panicles; flowers pale tangerine, the standard veined deep red; fruit a moniliform loment of 4–6 articles, these glabrous at maturity although often appressed-pubescent when juvenile.

Grassy campo.

Numerous moniliform fruits borne in large panicles give this species a very characteristic appearance. Widely dispersed throughout Central and South America; the type was collected in Bahia, Brazil by Sellow.

Estrada from base-camp to Rio Urariquera, *Lewis* 1587 (fl.)!

Aeschynomene sensitiva Sw., Prodr. Veg. Ind. Occ.: 107 (1788).
Only the typical variety occurs on the island.

var. **sensitiva** Fig. 6P.

Hollow-stemmed herb to shrublet, 1.5–2 m, characteristically drying blackish, stems glabrous, longitudinally ribbed; stipules not striate, appendaged below their point of insertion, acuminate; leaves pinnate, the petioles clothed with glandular-based hairs; inflorescence an axillary raceme, peduncle as for petioles; flowers yellow, the standard veined red; fruit a loment of 4–12 articles, these glabrous to sparingly hispid.

Growing in water of trackside swamp and in damp campo.

Widely distributed in wet places from the West Indies through Central and South America and "in the Old World tropics, apparently as an introduction" (Rudd, 1955).

Estrada from base-camp to the Rio Urariquera, *Lewis* 1428 (fr.)!; same locality, *Lewis* 1475 (fl. & fr.)!; in front of estação, *Hopkins et al.* 888 (fl.)!; same locality, *Ratter et al.* 5579 (fl. & fr.)!

51. **SOEMMERINGIA** Mart.

1 sp. in Brazil, Venezuela & Bolivia.

Soemmeringia semperflorens Mart., Diss. de Soemm.: 28 (1828). Fig. 6G, Plate 8C.

A prostrate, creeping herb from a central tap root; stipules linear-lanceolate, sharply acuminate, striate, papery, persistent; leaves pinnate with 3 pairs of leaflets, sensitive; leaflets bluish green, oblanceolate to obovate, shallowly emarginate at apex, glabrous, the blades asymmetric about the midvein, the secondary venation parallel and prominent on both faces, extending to the leaflet margin and forming small teeth; flowers solitary in the leaf axils; bracteoles ovate, persistent at calyx base; calyx green, the teeth much longer than the tube, the upper two broader and fused almost to their apex; corolla yellow fading to a pale lemon yellow, the standard becoming brownish and scarious and persisting as a delicately veined protective 'case' completely enclosing the green plicate fruit.

Open waste ground.

Near base-camp, *Lewis* 1600 (fl. & fr.)!

52. **ZORNIA** J.F. Gmel.

Mohlenbrock in Webbia 16(1): 1–141 (1961) (monograph of the genus, including a key to species and distribution maps) and in Webbia 16(2): 649–655 (1962).

About 80 spp. in tropical and warm temperate areas of the Old and New Worlds. Brazilian species currently under revision by Brandão Ferreira.

Zornia sp. aff. latifolia Sm. in Rees, Cycl. 39: 4 (1819). Fig. 6T.

Weedy herb to about 75 cm; stipules appendaged below their point of attachment, the main body of the stipule ovate-lanceolate, 5–6-veined, the appendage narrower, 3-veined; leaves consistently bifoliolate; leaflets ovate to elliptic or lanceolate, appressed-pubescent on both faces, the lower face with scattered punctate glands; inflorescence an axillary spike, the flowers sessile or almost so, the characteristic foliaceous bracts similar to the stipules but larger, the appendage relatively shorter, the venation more branched and the blade with punctate glands; flowers mostly enclosed within bracts; petals pale yellow; fruit a loment of ca. 4–6 articles, the articles with stiff bristles, both the article body and the bristles with a fine pubescence.

Waste ground.

Zornia latifolia falls into a group of species that need detailed revision. The collection from Maracá differs from typical material in possessing punctate glands on the foliage and bracts, characteristics suggesting affinities with *Z. glabra*.

Base-camp, *Lewis* 1599 (fl. & fr.)!

53. **STYLOSANTHES** Swartz

Mohlenbrock in Ann. Miss. Bot. Gard. 44(4): 299–355 (1957) (revision of the genus, including a key to species); 't Mannetje in Austr. Journ. Bot. 25(3): 347–362 (1977); Brandão Ferreira & Sousa Costa, 'O gênero *Stylosanthes* no Brasil', EPIMAG, 108 pp. (1979) (includes a key to Brazilian species); Stace & Edye, The Biology and Agronomy of *Stylosanthes,* Academic Press, xvii + 636 pp. (1984).

About 25–30 spp. in the tropics and warm temperate areas of the Old and New World.

Stylosanthes humilis Kunth, Nov. Gen. & Sp. Pl. 6: 506 (1823). Fig. 6S.

A prostrate to semi-erect herb; stipules scarious, fused to each other and to the basal part of the petiole but with free filiform apices, sparsely glandular-pubescent, persistent even after leaf fall; leaves pinnately trifoliolate; leaflets bluish green, narrowly elliptic to linear-lanceolate, glabrous but for a few scattered, yellowish glandular hairs; inflorescence an axillary or subterminal, few-flowered, congested head, the bracts and bracteoles with a dense clothing of long, stiff, yellowish, glandular hairs; corolla yellow; fruit usually a single 1-seeded article with the style persisting as an apical coiled beak.

Waste ground.

Near base-camp, *Lewis* 1694 (fl.)!

Stylosanthes guianensis (see Fig. 6R) collected within a few kilometres of the island, along the estrada to Boa Vista, has not been collected on the island but may ultimately appear in the disturbed ground near base-camp.

54. **CROTALARIA** L.

Windler and Skinner have prepared an account for Flora Neotropica.

About 600 species in the tropics and subtropics; most numerous in tropical Africa. 71 species in the Americas.

1. Leaves trifoliolate; inflorescence terminating in a persistent, dense cluster of curled, silky-pubescent bracts, these 5–15 mm long ... ... **C. micans**
1. Leaves unifoliolate; inflorescence not terminating as above:
   2. Stipules lacking and stems not winged between the nodes; inflorescence few-flowered ... ... **C. sagittalis agg.**
   2. Stipules decurrent, the stems thus with prominent wings between the nodes:
      3. Leaflets linear to narrowly lanceolate; stipules not becoming broader towards apex, free portion straight, ca. 5 mm long ... ... **C. pilosa**
      3. Leaflets obovate; stipules wing-like, becoming broader towards apex, free portion falcate, 10–25 mm long ... ... **C. stipularia**

Crotalaria micans Link, Enum. Hort. Berol. Alt. Part 2: 228 (1822). Fig. 6F.
Crotalaria anagyroides Kunth (1824).

Woody-based perennial herb or shrublet to 2.5 m; young stems with an appressed silky indumentum; leaves digitately trifoliolate; leaflets discolourous, upper face mid-green, lower face whitish green; inflorescence a terminal or, less often, axillary raceme characteristically terminated by a cluster of curled bracts, these falling as flowers reach anthesis; petals yellow, the keel markedly beaked; fruit a stipitate, inflated, dehiscent pod, (3–)4(–5) cm long, drying blackish; seeds yellowish ochre, kidney-shaped.

Trackside scrub or capoeira, sometimes in water.

As in other species of the genus, mature seeds are released from the placenta prior to pod dehiscence thus causing the fruits to rattle in the breeze; the generic name, *Crotalaria*, or rattle box alludes to this characteristic. Perhaps introduced onto the island by man in the last 10 years. *Milliken* 340 records large black bees [probably a species of *Xylocopa*] visiting the flowers.

Estrada from base-camp to Rio Urariquera, *Lewis* 1572 (fr.)!; same general locality, *Lewis* 1692 (fl. & fr.)!; near base-camp, *Hopkins et al.* 788 (fl.)!; track between base-camp and river, *Milliken* 340 (fl. & fr.)!

Crotalaria pilosa Miller, Gard. Dict. ed. 8, n. 2 (1768). Plate 8F.

Herb to 1.5 m; stems prominently winged by decurrent stipules, these not separating into free lobes at the leaflet base, bluish green with a dense silvery appressed indumentum; leaves unifoliolate, linear-oblong; inflorescence a terminal or axillary raceme, up to 40 cm long; small, 2–3 mm long, bracts regularly persisting after the flowers have fallen; flowers pale yellow; fruit an inflated, dehiscent pod, (3–)4 cm long; seeds brownish.

Damp grassy campo and seasonally flooded savanna.

By base-camp, *Lewis* 1464 (fl. & fr.)!; same locality, *Lima* 789 (fl. & fr.)!; same locality, *Milliken* 215 (fl. & fr.)!; same locality, *Ratter et al.* 5388 (fl.)!

Crotalaria sagittalis L., Sp. Pl. : 714 (1753), *sensu lato.*

Erect herb to ca. 50 cm, whole plant densely spreading pubescent with silvery or white hairs; stipules lacking; leaves unifoliolate, subsessile, bluish green, linear-lanceolate; flowers solitary or in few-flowered axillary racemes; bracts ± persistent; petals yellow; fruit an inflated, dehiscent pod, ca. 2.5 cm long.

Campo; damp grassy savanna.

Locally common in the grassy campo near base-camp and initially thought (by G.P.L.) to represent a new taxon. Material sent to Windler at Towson, Maryland, U.S.A. prompted the following comments: "the plant falls within the group of extremely variable relatives of *C. sagittalis* L. which come down from the north .... size seems not to be important — I have grown a 3 ft. plant from a 3 inch parent. Some populations have stipules and others do not. It is truly a mess".

Near base-camp, *Lewis* 1481 (fl. & fr.)!; same locality, *Milliken* 418 (fl. & fr.)!

Crotalaria stipularia Desv., J. Bot. (Desvaux) 3: 76 (1814).

Herb to 50 cm; whole plant with a dense, appressed to spreading, white to silvery indumentum; stems winged by decurrent, foliaceous stipules, these widening apically into 2 free, incurved lobes; leaves unifoliolate, sessile, glaucous bluish green; inflorescence an axillary raceme with persistent bracts; flowers yellow; fruit an inflated dehiscent pod.

Man-made soil mounds.

Probably introduced.

Near base-camp, *Lewis* 1693 (fl. & fr.)!

REFERENCES

ALTSCHUL, S. von REIS, A taxonomic study of the genus *Anadenanthera*, Contribs. Gray Herb. Univ. Harvard 193: 1–65 (1964).

—— The genus *Anadenanthera* in Amerindian cultures, Bot. Mus. Harvard Univ., Cambridge, Mass. 96 pp. (1972).

AMSHOFF, G.J.H., On South American Pipilionaceae, Utrecht 78 pp. (1939).

AUBRÉVILLE, A., Etude écologique des principales formations végétales du Brésil et contribution à la connaissance des forêts de l'amazonie brésilienne, 268 pp. (1961).

AZEVEDO, A.M.G. de, O gênero *Desmodium* Desv. no Brasil: Considerações Taxonômicas, Dissertação apresentada do Instituto de Biologia da Universidade Estadual de Campinas: 315 pp. (1981).

BENTHAM, G., Notes on Mimoseae, with a synopsis of species, Tribe III Acacieae (cont. from Vol. 3, p. 226), XXVIII *Inga* Martius, Hook., Lond. Journ. Bot. 4: 577–622 (1845).

—— A synopsis of the Dalbergieae, a tribe of the Leguminosae, Journ. Linn. Soc. 4, Suppl.: 1–134 (1860).

—— VII Revision of the Suborder Mimoseae [extracted from the 'Transactions of the Linnean Society of London'] 30: 335–664 (1875).

COWAN, R.S., A taxonomic revision of the genus *Macrolobium* (Leg. — Caesalpinioideae), Mem. N.Y. Bot. Gard. 8(4): 257–342 (1953).

—— *Swartzia* (Leg. — Caesalpinioideae, Swartzieae), Flora Neotropica, Monograph No. 1, 228 pp. (1968).

—— A taxonomic revision of *Elizabetha* (Leguminosae — Caesalpinioideae), Konikl. Nederl. Akad. Wetensch. Amster., series C, 79(4): 323–346 (1976).

DUCKE, A., Plantes nouvelles ou peu connues de la region Amazonienne, Archivos. Jard. Bot. Rio de Janeiro 3: 47–175 (Leguminosae) (1922).

—— As leguminosas do estado do Pará, Archivos. Jard. Bot. Rio de Janeiro 4: 211–341 (1925) and 5: 191–199 (1930).

—— & BLACK, G.A., Phytogeographical Notes on the Brazilian Amazon, An. Acad. Bras. de Ciencias 25(1): 1–46 (1953).

———— Notas Sobre a fitogeografia dá Amazonia Brasileira, Bol. Técn. do Inst. Agron. do Norte 29: 1–62 (1954).

FEVEREIRO, V.P.B., *Centrosema* do Brasil, Rodriguésia 29(42): 159–219 (1977).

—— *Macroptilium* do Brasil, dissertação de mestrado, Rio de Janeiro: 160 pp. (1979).

GREAR, J.W. Jr., A revision of the American species of *Eriosema* (Leg. — Lotoideae), Mem. N.Y. Bot. Gard. 20(3): 1–98 (1970).

—— A revision of the new world species of *Rhynchosia*, Mem. N.Y. Bot. Gard. 31(1): 1–168 (1978).

HEMMING, J.H.(ed.), Change in the Amazon Basin, Vol. 1: Man's impact on forests and rivers, Manchester University Press: ix + 222 pp. (1985).

HOPKINS, H.C.F., *Parkia* (Leguminosae — Mimosoideae), Flora Neotropica, Monograph 43, 124 pp. (1986).

IRWIN, H.S., Contributions to the Botany of Guiana. III Leguminosae — Mimosoideae, Mem. N.Y. Bot. Gard. 15: 96–111 (1966).

—— Contributions to the Botany of Guiana IV Leguminosae — Caesalpinioideae, Mem. N.Y. Bot. Gard. 15: 112–128 (1966).

—— & BARNEBY, R.C., The American Cassiinae: a synoptical revision of Leguminosae Tribe Cassieae subtribe Cassiinae in the New World, Mem. N.Y. Bot. Gard. 35 (parts I & II): 1–918 (1982).

JACKSON, B.D., A Glossary of Botanic Terms with their derivation and accent, 4th ed., 481 pp. (1971).

LAVIN, M., Systematics of *Coursetia* (Leguminosae — Papilionoideae), Systematic Botany Monographs, Vol. 21, 167 pp. (1988).
LEE, Y.T. & LANGENHEIM, J.H., New taxa from Brazil & Guyana in the genus *Hymenaea* (Leg. Caesalpiniodeae), Journ. Arn. Arbor. 54(1): 94–104 (1973).
—— Additional new taxa and new combinations in *Hymenaea* (Leg. Caesalpinioideae), Journ. Arn. Arbor. 55(3): 441–452 (1974).
—— Systematics of the genus *Hymenaea* L. (Leg. Caesalpinioideae, Detarieae), Univ. of Calif. Publications in Botany 69: 1–109 (1975).
LEWIS, G.P., Legumes of Bahia, Royal Botanic Gardens, Kew xvi + 369 pp. (1987).
LIMA, H.C. de, Considerações taxonômicas sobre o gênero *Hymenolobium* Benth. (Leg. Fab.), Acta Amazonica 12(1): 41–48 (1982).
LÖFGREN, A., "Plantes nouvelles ou peu connues de la région Amazonienne", Archivos. Jard. Bot. Rio de Janeiro 3: 1–281 (1922).
MACBRIDE, J.F., Flora of Peru, Leguminosae, Publications Field Mus. Nat. Hist. Chicago, Bot. Series 8(3): 507 pp. (1943).
MARÉCHAL, R., MASCHERPA, J.-M. & STAINIER, F., Etude taxonomique d'un groupe complexe d'espèces des genres *Phaseolus* et *Vigna*, Boissiera 28: 1–273 (1978).
MARTIUS, C.F.P. von, Flora Brasiliensis, Leguminosae 1, Papilionaceae, 15(1): 1–350 (1859–1862) & Leguminosae 2 & 3, Swartzieae, Caesalpineae, Mimoseae 15(2): 1–527 (1870–1876).
MATTOS, N.F., O gênero *Andira* (Leg. Pap.) no Brasil, Acta Amazonica 9(2): 241–266 (1979).
MAXWELL, R.H., The genus *Dioclea* (Fabaceae) in the New World, Dept. of Bot., Southern Illinois University: 430 pp. (1969).
MELLO FILHO, L.E. de & ANDRADE, A.G. de, Espécies amazônicas do gênero *Zollernia* (Caesalp.), Atas do Simpósio sôbre a biota amazônica 4 (Botânica): 153–166 (1967).
MOHLENBROCK, R.H., A revision of the genus *Stylosanthes*, Ann. Miss. Bot. Gard. 44(4): 299–355 (1957).
—— A monograph of the Leguminous genus *Zornia*, Webbia 16(1): 1–141 (1961).
MOSKOVITS, D.K., The behaviour and ecology of the two Amazonian tortoises, *Geochelone carbonaria* and *G. denticulata* in northwestern Brasil, Dept. of Biology, Chicago, Illinois (1985).
NASH, R.J. et al., Castanospermine in *Alexa* species, Phytochemistry 27(5): 1403–1404 (1988).
PITTIER, H., Preliminary revision of the genus *Inga*, Contrib. U.S. Nat. Herb. 18(5): 173–223 (1916).
POLHILL, R.M. & RAVEN, P.H. (eds.), Advances in Legume Systematics Part I, xvi + 425 pp. (1981).
PONCY, O., Le genre *Inga* (Leg., Mim.) en Guyane Française, Mém. Mus. Nat. d'Hist. Nat. Paris, nouvelle série, 31: 1–124 (1985).
PRANCE, G.T. & LOVEJOY, T.E., Key Environments: Amazonia, Pergamon Press, Oxford: xiv + 442 pp. (1985).
RAMIREZ, N.L., Revision taxonomica del genero *Alexa* Moq. (Leguminosae: Papilionoideae, Sophoreae) en Venezuela, Facutad de Ciencias, Universidad de Venezuela, 67 pp. (1983).
RICE, A.H., The Rio Branco, Uraricoera and Parima, Geographical Journal 71: 113–143, 209–223, 345–357 (1928).
ROOSMALEN, M.G.M. van, Fruits of the Guianan Flora, Utrecht 483 pp. (1985).
RUDD, V.E., Review of the Genus *Centrolobium*, Journ. Wash. Acad. Sci. 44(9): 284–288 (1954).

—— The American species of *Aeschynomene,* Contrib. U.S. Nat. Herb. 32(1): 1–172 (1955).
—— Supplementary studies in *Aeschynomene,* I: Series Viscidulae, including a new species and five new varieties, Journ. Wash. Acad. Sci. 49(2): 45–52 (1959).
—— The American species of *Ormosia* (Leg.)., Contrib. U.S. Nat. Herb. 32(5): 279–384 (1965).
—— Studies in *Coursetia* (Leg.) I, Phytologia 25(4): 257–258 (1973).
—— New taxa and combinations in *Machaerium* (Leg.) III, Phytologia 25(6): 398–403 (1973).
—— New taxa & combinations in *Machaerium* IV, Phytologia 26(2): 100–102 (1973).
SANDWITH, N.Y., Contributions to the Flora of Tropical America VII, New and noteworthy Leguminosae and Rosaceae from British Guiana, Kew Bull. 1931: 357–377 (1931).
SAUER, J., A Revision of *Canavalia,* Brittonia 16: 106–181 (1964).
SCHOMBURGK, R., Richard Schomburgk's Travels in British Guiana, 1840–1844, translated by W.E. Roth, "Daily Chronicle" Office, Georgetown, 1: xxxviii + 402 pp. (1922) & 2: 443 pp. (1923).
SCHOMBURGK, R.H., Travels in Guiana and on the Orinoco during the years 1835–1839, translated by W.E. Roth, "The Argosy" Co. Ltd., Georgetown, British Guiana: viii + 202 pp. (1931).
SEMA, Programa de estações ecológicas, Brasília: 40 pp. (1977).
SILVA, M.F. da, Revisão taxonômica do gênero *Peltogyne* Vog. (Leg. Caesalp.), Acta Amazonica 6(1): 1–61 (1976).
—— Nomes vulgares de plantas amazônicas, CNPq/INPA, Manaus–Amazonas, 222 pp. (1977).
STACE, H.M. & EDYE, L.A., The Biology and Agronomy of *Stylosanthes,* Academic Press, xvii + 636 pp. (1984).
TROPICAL LEGUMES: Resources for the Future, National Academy of Sciences, Washington DC., x + 331 pp. (1979).
VERDCOURT, B., A Manual of New Guinea Legumes, 645 pp. (1979).
WINDLER, D.R. & SKINNER, S.G., New taxa & new combinations in the American Crotalarias (Fab.), Phytologia 50(3): 185–206 (1982).
YAKOVLEV, G.P., A review of *Sweetia & Acosmium,* Notes from R.B.G. Edinburgh 29: 347–355 (1969).
—— Survey of Genera, *Zollernia* Nees and *Lecointea* Ducke, Botanicheskii Zhurnal 61(9): 1304–1308 (1976).

INDEX TO SCIENTIFIC NAMES

Names in brackets are synonyms. Species or genera not recorded from Maracá but mentioned in the text are preceded by a short line. Reference to figures in bold type.

INDEX TO VERNACULAR NAMES